MANAGING NEW TECHNOLOGICAL CHANGE

For Anastasia

Managing New Technological Change

Case studies in the reorganization of work

PETER WILKINS

Avebury

Aldershot · Brookfield USA · Hong Kong · Singapore · Sydney

Published by
Avebury
Ashgate Publishing Limited
Gower House
Croft Road
Aldershot
Hants GU11 3HR

Ashgate Publishing Company
Old Post Road
Brookfield
Vermont 05036
USA

A CIP catalogue record for this book is available from the British Library of Congress.

ISBN 1 85628 336 4

Printed and Bound in Great Britain by
Athenaeum Press Ltd., Newcastle upon Tyne.

Contents

Acknowledgements

This book is the result of a project conducted over three years and financed by a joint committee of the ESRC/SERC.

There are a number of people whose assistance I would like to acknowledge. The first of these are those companies and individuals who took part in the study. Some were undoubtedly more willing than others, but, with very few exceptions, nearly all were accommodating and helpful.

I received a great deal of assistance and advice from Bryn Jones of the University of Bath. Peter Cressey, also at Bath, and Graeme Salaman, of the Open University, supplied further useful commentaries of the original transcript of the work.

I owe a considerable personal debt to my wife Anastasia Paris for her support of this project and her significant practical assistance. Abundant personal support was also received both from my parents, Heather and Harold Wilkins, and parents-in-law Mary and Gregory Paris. Further practical assistance was given by Professor Prodromos Efthymologou.

The book that results, and any errors it may contain, remains, of course, entirely my own responsibility.

1 Introduction

The problem

New technology changes jobs but how are these jobs defined?

Job definition and work organization is assumed by technological determinists[1] to be directly informed by the kind of technology employed and therefore unproblematic. Woodward's (1966) work, for example, asserted that management organization varied according to the nature of manufacture, whether single units or large or small batches. Similarly, Blauner (1964) considered that different forms of technology offered different forms of control. Recognizing that social and political factors were involved in the process Braverman (1974) nonetheless still maintains that there is one direction of cause and effect associated with new technology introduction; in this case it is deskilling for managerial control based on economic determinism.

These uni-directional approaches are widely disputed by many others who argue that there is choice in technology and work design (Littler and Salaman, 1984; Davis and Taylor, 1976; Wilkinson, 1983).[2] These choices are influenced by a range of unique factors (Buchanan and Boddy, 1986) some of which are extra-organizational (Salaman, 1986). Thus, even two identical technologies may ostensibly result in quite different organizations of work.

One problem in managing work reorganization stems from this choice of new technology and design of jobs. The key question that emerges from here then is: what are the processes in new technology introduction that inform the way jobs are designed and who are the actors capable of influencing the process? Given that the process involves a mixture of technical, social and political variables the use of new technology will fluctuate according to these forces. This variability in the decision making process itself creates major dilemmas and debates in attempts to arrive at the most suitable job definitions and work organizations.

Various decision making processes are involved: from the initial implementation through to decisions about work organization and who will fill the individual jobs. The key actors are the managers vested with the authority to make the decisions, the workers directly affected by the decisions and the trade unions perhaps with the power to influence the outcomes. But to what extent do managers recognize the existence of, and how much do they exercise choice? How much direct influence are workers able to have? How far do trade unions become involved in the job definition process? These are some of the questions that this book addresses in the following chapters.

Accounts already exist that have tackled issues relating to choice and job definition such as the sociotechnical approach. Another account (Child, 1972) has suggested that a range of choices are often available but failure on the part of managers to consider these means they choose '....unconsciously in blind response to immediate pressures' (Rose 1988, p346). However, there is an important connection to be made between job definition and the various processes of decision taking that emerge when new technology is introduced. Technological determinism and economic determinism are inadequate models for explaining these processes. Yet in pilot studies managers regularly referred to the lack of choice available in work organization once the new technology itself was chosen, they assumed the work organization issues to be unproblematic. However, contradicting this deterministic view, criteria distinctly socio-political in nature were being applied in job definition and allocation decisions. This contradiction forms a central plank of the study which aims to identify and explain some of the job definition aspects of work organization issues arising from new technological change.

The approach[3]

Defining new technology as microelectronic systems for the processing and communication of information and control, for the purposes of the study, preliminary pilot studies were set up. Following these, six companies were chosen for detailed analysis of work reorganization and job definition issues.

The pilot studies were carried out on a random selection of companies found through press reports, local chambers of commerce, engineering publications and help from researchers already working in the area. They included companies in different sectors and employing very different forms of new technology.

Simultaneously, interviews were carried out with officials from trade unions such as the AEU, TASS (now amalgamated with the ASTMS, and called MSF), APEX, NCU, NGA and SOGAT (the NGA and SOGAT are now amalgamated and called GPMU).[4]

Based on the information gained from the pilot stage, six companies were targeted and indepth interviewing commenced. A considerable range of possibilities existed in the choice of companies. One option was to concentrate on one set of companies operating the same technologies. It was possible also, to either use in-depth interviewing in relatively few firms, or to conduct a small number of interviews in many more companies.

Attempting to recognize both the need for comparability and for generalization, a broader selection of technologies was combined with a narrow selection of firms. The full complexity of job definition issues and decision making, it was believed, would be difficult to investigate from the limited data of a few interviews.

The chosen sectors, metalworking, printing and offices, all have recently experienced significant technological change but the industrial and commercial sectors they represent are all very different in the kinds of factors that inform and influence them. Some have strong unions. In some the introduction of new technology has had a more significant impact than in others. This gives some basis for generalization should there be common findings across these three, quite different, areas. Moreover, in each case the companies are each matched with another firm from the same comparable sectors. Each pair of companies has the same trade unions and similar technologies. They are also broadly comparable in the nature of the work they do. These matched pairs of companies complement each other and act as controls on each other, thereby reducing the possibility of special or atypical results.

Choosing only six companies also meant that a fairly indepth study of each was possible. Since this book is about the management of changes in work organization, the key actors are the decision makers and those capable of influencing the decisions. In every case the interviews started with senior managers and then worked down the hierarchy. This was intended to give an overall perspective before focusing on to the local areas where the job changes were actually occurring. Subsequent interviews involved supervisors, various grades of worker, and trade union representatives. Hence a range of views were collected from those people in the company who were affected by, or may have some effect on, the new technology change.

Organization of the book

The book is arranged in eight chapters. Five chapters relate solely to the case study analysis and description. In addition to the overall introduction and conclusion, the remainder deal with the theoretical issues relevant to the project, and the analysis of existing case study material.

In chapter two theories of job design are related to complementary theories of technological change, and this in turn is related to industrial relations issues. The early theorists of the division of labour Smith (1776), Babbage (1835) and Taylor (1947), all argue for a highly divided form of job design and more recent writers in the labour process tradition believe that there is a tendency for managers to organize work in this way (Braverman, 1974). However, alternatives are offered in the sociotechnical systems approach. In other studies there is some evidence of jobs being organized to include more tasks either of the same status level or of a higher status than those existing in the job (see Wood, 1989).

The chapter moves on to consider the choice available between different forms of technology and the extent to which the introduction of this has been planned. The role of trade unions, and more general industrial relations issues, are considered as contributors to the reorganization of work roles. The trends toward flexible work organization and its relationship to new technology are investigated and discussed in the final parts of the chapter.

Findings from existing case studies corresponding with the new technologies chosen for analysis are considered in chapter three. The purpose of this chapter is to take up the main theoretical issues that are raised in chapter two. The data from each study is considered under three main headings: managerial intentions, work organization, trade union involvement. Managerial intentions is a consideration of the planning that goes into the technology and job change issues. Work organization looks at the changing job definitions and problems that have occurred as a result of these, and trade union involvement considers the extent of trade union influence in technology and job design issues. Finally, in chapter three, five hypotheses are derived from the review of the existing literature and case studies.

Descriptive background data on the six case study companies, presented in chapter four, provides information that helps to put the job changes, discussed in later chapters, in context. These later chapters (five, six and seven) discuss the problems of the nature of work, work allocation, planning for technological change, trade union involvement and occupational mergers.

Accounts of new technology's effect on jobs polarize around the issue of whether there is an increase or a decrease in the skill content of jobs. The first part of chapter five tackles this issue. Moreover, as a discussion of work organization in the case study companies, it also provides essential background data for the understanding of the job changes in the companies. The hypothesis claims that redesigned work will create both deskilled and reskilled work, generalized changes of either deskilling or reskilling are not taking place.

The literature on skill changes and technological reorganization of work contains many claims about the causes and character of new jobs. But rarely have the ways in which particular individuals or categories of these come to occupy the new jobs, the process of labour allocation, been explicitly examined. Yet this must have some influence on skill levels and task requirements. The second part of chapter five hypothesizes that organizational, social and political factors may lead to a different kind of worker being placed in reorganized jobs than technical criteria alone would suggest.

In literature on new technology introduction (Buchanan, 1986; Buchanan and Boddy, 1986; Child, 1972; Rose and Jones, 1985), key decisions are taken by managers, from the decision of which technology to employ, through to the designing of jobs. However, it remains unclear what patterns this procedure takes. In chapter six this question is addressed in three parts. The first looks at managerial planning for change, the second at the reasons managers attribute to the changes, and the third considers who was involved in the decisions. The hypothesis emerging from the analysis of existing literature and case studies states that strategy for change is rare, particularly in respect of job design. Rationales for change are often vague and fail to reflect general motives of control. Finally, discussion for the formulation of decisions that does take place is often confined to managerial grades.

Various accounts claim that trade unions are effective in influencing new technology changes. Often examples of union success in limiting redundancies, establishing redeployment terms and negotiating pay increases, for the operation of new technology, are cited. However, there is little consideration about trade union knowledge and ability to influence the way work is organized and jobs are designed. The first part of chapter six deals with the hypothesis that trade union knowledge of job design and work organization issues at local level is limited. As a result of this union ability to respond to challenges on these issues and make a case for inclusion in the job design debate, is restricted.

In the introduction of new technology, job boundaries often become blurred and this may result in a blurring of job definitions in which some workers will be called upon to take on new tasks which may have previously been within colleagues' job descriptions. The problem of union conflict that results has been identified by other writers, but rarely specifically tested. The second part of chapter seven deals with a new hypothesis that trade union conflicts over demarcation issues, both within and between unions, are increasingly common, due to new technology introduction. This potentially weakens the unity and power base of trade unions as collectives.

The conclusion forms chapter eight and attempts an overall summary of the book, drawing the main conclusions from the research findings. Confirming all of the hypotheses, to varying degrees, the study finds that there are a range of choices in new technology introduction that managers and trade union often fail to acknowledge. Technologically deterministic views mean that managers often see no need to plan work assuming that the technology automatically makes these decisions. However, there are real dilemmas that must be resolved in work organization and job definition such as the division of labour and the allocation of work. Failure to consider these issues means that decisions become subject to sets of *ad hoc* criteria that reflect social and political views as well as technical and economic ideas.

The book concludes by suggesting that this contradiction between technological determinism on the one hand, and the application of socio-political criteria on the other, may be resolved by careful planning that considers the choices available in work organization. Part of this process may be assisted by management, worker and trade union consultation that will offer opportunities for trade unions and workers to educate managers in available choice from their particular points of view.

Notes

1. Technological determinism is an approach that believes new technology systems themselves dictate the way that work is organized and allocated.
2. Although Noble (1979) argues from a labour process perspective that choice is available in machine design and the design preferences are shaped by capitalist notions of deskilling workers to gain greater control. In this way the argument returns to technological determinism.
3. See chapter four for details of each of the companies.
4. AEU - Associated Engineering Union (formerly AUEW - Associated Union of Engineering Workers); APEX - Association of Professional and Executive Staff; TASS - Technical and Supervisory Staff Association; ASTMS - Association of Scientific, Technical and Managerial Staff, MSF - Management Science and Finance Union; NGA - National Graphical Association; SOGAT - Society of Graphical and Allied Trades; GPMU - Graphical Paper and Media Union.

2 New technology, skill and the labour process

Introduction

Chapter one established that the objective of this book is to consider the factors that influence work organization following new technological change. The main issues of relevance concern, firstly, the identification of those capable of influencing the process of change, and the level of influence that they have. Secondly, the resulting work organization and the allocation of jobs that takes place.

The existing literature deals with some of these questions in studies of job design, skill, new technology and industrial relations, for example Hyman (1988) and Wood (1982; 1989). In this chapter other influences on the division of labour are considered including social and political factors that shape the choice of new technology as well as work organization. Industrial relations issues are the subject of the latter part of the chapter.

The chapter first considers different perspectives that explain why jobs are divided up as they are and the necessity or otherwise of this division of labour. Four perspectives are discussed:

1. The classical division of labour approach.
2. The sociotechnical perspective.
3. The labour process theory.
4. Theories of post Fordist trends in work.

The second part of the chapter introduces the new technology dimension which deals with three aspects of change introduced by new systems:

5. The nature of current technological changes.
6. Political and social factors in new technology and automation.
7. The distribution of skill and new technological change.

These sections deal with the decision to introduce new technology, the choice in the way it is used through to the implications of new technology on skills according to existing literature.

Finally, the chapter deals with industrial relations issues from managerial and trade union points of view. This questions managerial approaches to new technology, and the levels of influence of trade unions on new technology with particular reference to the work organization and design aspects. The last section looks at trends toward establishing a flexible workforce.

Classical traditions of job design

Adam Smith is perhaps considered the first to detail a management system involving a specialized division of labour. The system was characterized by three main principles. These are the increase of productivity due to enhanced dexterity, minimum change over times between operations and the decomposition of tasks, an aid to stimulating the development of new machines. (Littler, 1985 p11).

Charles Babbage (1835, p3-5), some sixty years later, argued that manufacturers are compelled to consider the cheapest form of production by competition they face he believed the most important principle was the division of labour. With these beliefs he built upon Adam Smith's conception of the division of labour adding that specialization would also reduce time in training and expenditure during the learning process. It would also exclude the need to use a variety of tools (considered to be time consumers), an extension of Smith's reducing change over times. Babbage's contribution, a demonstration of how the Smith process could increase profitability by reducing the cost of labour, is sometimes referred to as the 'Babbage Principle.' The essential thrust of both Smith's and Babbage's analyses were economic ones; dividing labour would serve to reduce training time and change over times, reducing costs thereby. The central inference from these accounts, fundamental to the attempt to divide up work, is the motive to gain control over the labour process. Organizing work in small detailed units negates the possibility of collective organization and enables managers to understand the functions of each worker. Frederick Taylor was quite specific about such a system and explicit about the control motive.

Smith and Babbage were the forebearers of Frederick Winslow Taylor's systematic style of management, described by him as 'Scientific Management' and by others as 'Taylorism.' Taylor might more accurately be described as a neo classical theorist because he elaborated something that neither Smith nor Babbage had presented in their analyses. Job specialization and narrowing could be used to divide mental from manual labour, this formed the very essence of the Taylor system.[1]

Scientific Management comprised a rationalization of labour and a systematic form of management. The emphasis was on the control of labour and not on the manufacturing process itself as occurred later with Fordism where the control processes are largely contained within the machinery.[2] Scientific management involved the design of jobs and control over task performance. Taylor maintained that the way to achieve the maximum efficiency was to find the best way of performing a task by analysis of the speed at which the task could be completed, and set targets for workers to attain, this he described as a 'task idea.'

Workers would not give their all, Taylor maintained, unless they were paid incentives, so an incentive system was built into Taylorism. It was felt that this would serve to motivate workers and prevent them from working under their ability.[3]

Perhaps one of the most important features to be found in Taylor's work is the criticism of existing management as ignorant of the tasks performed by workers. The significance of Taylor's attitude towards management reflected his idea of the need for a command structure. Too often workers were left to take control of work operations, and managers were displaying an impotence and inability to control because they were unaware of job content. He found evidence of this in the worker organized concept of 'soldiering' which describes the purposeful restriction of output by workers.

Taylor took a particular manual process and tended to examine the most efficient way that manual tasks could handle this; for example Schmidt's loading of pig iron (1911, p44ff).

The concept of Fordism is often linked to Taylorism in writings on the division of labour, however, the essential difference of Fordism is that Ford would not consider the best way of handling a manual task but would instead investigate mechanization of that task from the point of view of the production process as a whole.

While some writers have suggested classical job design methods are those currently applied, others have tended to shift the emphasis away from economic efficiency to ideas that, do not ignore economic efficiency but, also consider the human element in work rather than seeing people as synonymous with machines.

Job design and the sociotechnical approach

The theory of sociotechnical design is an influential perspective that draws together a number of revisionist approaches based on the humanization rather than the economic efficiency of job design. It is therefore, an approach that is in opposition to Taylorism and Fordism. Unlike labour process and neo Fordist approaches (see later in this chapter), it actually has confidence in a system of design that incorporates an effective technological design, with a work design that benefits both the company and workers.

Critics suggest that this presents a somewhat idealized picture of the necessary route toward the redesign of jobs. Whilst the notion of designing jobs for worker involvement would meet with broad acceptance, the apolitical approach makes the schema problematic. There is no mention of a labour process view in terms of managerial control motives. However, the school does provide a learned account that is based on practical experience and empirical research. Its contribution to the job design debate is an important one and it addresses the precise issues with which this book is concerned: the choice of technology, the choice of job design and what informs these decisions.

The often referred to study of the Longwall Method of coal getting (Trist and Bamforth, 1951) provides a good example of work in the sociotechnical school. The study began by considering the organization of pre-longwall methods. This had involved manual extraction of coal and a good deal of worker autonomy with workers organized in self selecting teams operating across a broad range of tasks; they were even paid on a group basis. The longwall method, so called because it was a mechanized system, and enabled attacking a whole face, involved a detailed division of labour.

Instead of the workers performing a full range of coal getting tasks in small teams that operated across three shifts, the work was divided and individualized. Workers now undertook only one task according to the shift they worked. In a twenty four hour period the coal getting cycle would be complete, in theory. However, the failure of one shift to complete its work meant that the following shift could not do its work. Quite apart from this, absenteeism and sickness levels increased markedly, hostile management worker relations resulted over resentment of external control, and output fell.

The response to the problems was to experiment with an alternative form of organization that in many ways mirrored the work design use in the days of hand getting. The composite teams that emerged involved an expansion of the skill repertoire, self selecting work groups, and a bonus that would be paid to the work group, not to individuals. Comparison between composite groups and longwall organization revealed a lower absenteeism rate, maintenance of cycle targets, better worker management relations and improved worker satisfaction.

The emphasis of this study, and later, of sociotechnical theories generally, is that the whole system must be considered because the technical, social and economic dimensions all operate interdependently. While each of these may set limits to the overall organization there is choice within this framework and to claim that the technology dictates the work definition is quite false. Davis (1979b p33) has described such technological determinism as 'dangerously simplistic.'

The more cynical would view the wider implications of sociotechnical theory in much the same way that piecemeal attempts of reorganizing work, such as job enlargement and job rotation, are viewed. Writers in the Marxist labour process and neo Fordist traditions have considerable scepticism about such approaches believing they conceal control motives. However, the essential difference of rotation or enlargement of jobs and sociotechnical designs are ones of coverage. The former deal with small parts of, or modifications to, the organizational system aimed specifically at jobs. Sociotechnical designs are aimed at redesigning the whole system. Hence the name, encompassing both the social and technical. However, unlike other methods that have been examined, and placed within a neo Fordist framework, sociotechnical design claims for itself a genuine aim of achieving participation and control for the employee:

> Sociotechnical design has a clear ethical principle associated with it. This is to increase the ability of the individual to participate in decision taking and in this way to enable him or her to exercise a degree of control over the immediate work environment. (Mumford, 1986 pp334-335)

Here the Tavistock Institute of Human Relations' advocacy of autonomous work groups, offering a range of qualitative benefits to the workers, and conceived as an end in itself, questions the necessary appeal to a profit motive (Rose, 1988 p243).

Much of the work specifically considering job design has been undertaken by Louis Davis in the USA. Working with Canter in 1955, Davis attempted a definition of the concept. They saw it as 'the organization (or structure) of a job to satisfy the technical organizational requirements of the work to be accomplished and human requirements of the person performing the work' (Davis and Canter in Davis, 1979b p30).

Job design is thus embodied in three activities: (i) Specifying control of individual tasks. (ii) Specifying methods of performance for each task including machinery and tools. (iii) Combining individual tasks into specific jobs. It was recognized that this meant encountering several problems, including flexibility in design to meet technological change and worker resistance to change on the grounds of job security (Davis, 1979b). There is also a suggestion that technology will have a necessarily beneficial effect. In stark contrast to labour process writers, Davis and Taylor (1979) claim that the detailed division of labour which emerged with mass production, has been taken a stage further in which the technology is able to deal with most of the menial work. The work that is left requiring a higher standard of competence. This bears comparison with the work of Hirschhorn (1984) and Blauner (1964), and presents an argument that could itself be taken as a form of technological determinism.

Sociotechnical theory presents a particular view based on the successful design of work. However, questions of the extent to which it is employed, and the extent to which it is effective can only be answered by examining the evidence. Through an analysis of new and existing case studies current methods of job design are examined in the following chapters.

The labour process interpretation

From a labour process perspective, managers in a capitalist system are interested only in attempting to gain control over the workforce, in order to maximize their economic gains. The control is gained through the medium of different labour processes such as Taylorist or Fordist approaches to the organization and management of work. The predominant form of control is by reducing the skills of workers and expanding the knowledge of managers. In this approach all managers are driven by a motive of control because of the very nature of capitalism. The redefinition of the content and organization of work is required to achieve this control.

Braverman's (1974) work has become a focal point for recent theories of the Marxist labour process. In his book *Labor and Monopoly Capital* Taylorism or scientific management is defined as the model method and style of management that establishes the maximum control over the workforce. New technology plays an important role in enabling a deskilling or simplifying of work in the way that Taylor (1911) recommended in turn allowing the imposition of control through the medium of thorough managerial knowledge of tasks performed. At the same time as asserting ideas of deskilling for control, he rejects human relations and industrial psychology approaches as providing a basis for alternative forms of worker satisfaction and work organization. These being no more than: '....fig leaves which cover up the continuing application and institutionalization of Taylorist principles' (Coombs, 1978 p84).

Braverman is also dismissive of job enlargement schemes since his argument concludes that there is no escaping Taylorist philosophy within a capitalist regime. However, some important work in this area has demonstrated that a skilful combination of production operations, to organize along more humanistic lines, can deliver both an ideological advantage as well as an economic one (Coombs, 1978 p84).

There is a deviation in Braverman's version from a technologically deterministic view. Nevertheless, traces of this remain in the overall account, manifest particularly in a preoccupation with the notion that technology will be employed in one way only and this is unilaterally determined by managers. Nevertheless, Braverman does not subscribe to the view that engineers are the sole protagonists in technical matters:

> Rejecting the conventional prejudice amongst Marxists that the technical aspects of machinery should be left to engineers and empiricists, Braverman correctly insists that machinery should be subjected to a social as well as technical analysis (Coombs, 1978 p85).

Indeed, from a similar position to the labour process school Edwards (1978) also confirms the social dimension, arguing that the technology itself does not necessarily mean the worker will lose control over the workplace, but where it does '....this consequence must nearly always be understood to be the result of the *particular* (capitalist) design of the technology' (p115: Emphasis in original).

This is supported with an analysis of numerically controlled machine tools in which Braverman argues that not only the particular type of technology is important but also the way it is used is specifically designed to enhance control over the work environment. Indeed, it is also maintained that an alternative application and organization of such machinery would not result in the deskilling so desired by capitalists.

While his assertion that social criteria are of considerable importance is not incorrect, Braverman's notion of an omnipotent management and an inactive working class is problematic. Others, Marxists amongst them, have acknowledged that there are many other factors at work that influence the design and operation of the technology itself, apart from strengthening managerial, and therefore capitalist, control over the labour process.

Using Taylorism as the classical example of capitalist management, and seeing the various work humanization initiatives as a sop, Braverman argues that various technological forms that have developed, are successful in their aim to deskill the workforce, and thereby force it into strategically weak positions. Critics however, have argued that a multitude of other factors act on the labour process and effectively deny any possibility of a systematic capitalist strategy from being realized. This erasure of crucial conflicts such as the opposition of trade unions is central to the arguments raised against Braverman (see Friedman, 1977; Littler and Salaman, 1982). As Elger (1985) points out:

> For Braverman the process of degradation of work and the disciplining effect of the reserve army of labour together appear to produce a virtually inert working class,

unable to pose any substantial problem for capitalism either within capitalism or beyond it (p25).

While Braverman fails to involve labour as key actors in the process, others more accurately prefer to consider labour factors, as well as organizational ones (Sorge *et. al.*, 1983).[4] It is also Braverman's romanticized and idealized notion of craftsmanship and the nature of craft control which is brought into question. For some writers (see for example Littler, 1982 p11; Coombs, 1978 p86n(12); Littler, 1978 p192), while Braverman holds Taylorism as responsible for the destruction of craft autonomy, it may be the case that this autonomy had already been significantly eroded prior to the onslaught of Taylorism.

Amongst the examples from manufacturing Braverman focuses specifically on numerically controlled machines (NC) which he claims have been both designed and deployed in ways that assert managerial control. But in the case of NC considerable empirical evidence (Jones, 1982; Batstone *et. al.*, 1987) presents a challenge to the thesis as a whole, as well as to a specific technology. What Bryn Jones is able to demonstrate is that the loss of several traditional craft skills, does not necessarily mean the total loss of skills, and therefore also loss of discretion and control. Jones found cases in which programming and related tasks, involving aspects of planning, were undertaken on the shopfloor and were not removed to the office. Secondly, in other examples it is clear that even where full programming tasks were not employed, there were instances of the amending of programs and of minor adjustments to existing programs. Given that this is the case there is unlikely to have been any universal managerial strategy that can be said to attempt to achieve control via the deskilling of the workforce. If this did exist it has clearly failed.

However, Braverman has not addressed the issue of areas in which it is clear that skills have been retained or even acquired. The strength of accounts like that of Jones (1982), written in response to *Labor and Monopoly Capital*, lie in their empirical nature. A feature which Braverman's work lacks.[5] However, it has more recently been argued (Armstrong, 1988) that such empirical evidence is not adequate grounds for criticism. Contrary to the impression gained by a great many writers, Armstrong suggests that Braverman makes reference not to a universal tendency to deskill but only to a general tendency. He illustrates this with a well known quotation from Braverman on the issue of deskilling and consequent reskilling:

....there is no question that from a practical standpoint, there is nothing to prevent the machining process under numerical control from remaining the province of the total craftsman. That this <u>almost never happens is due</u>, of course, to the opportunities the process offers for the destruction of craft and the cheapening of the resulting pieces of labor into which it is broken (Braverman, 1974 p199: Emphasis added by Armstrong, 1988 - quoted p148).

Armstrong is arguing for a longer term analysis, claiming that isolated case studies give only snapshot views which are bound to reveal inconsistencies with the deskilling hypothesis. It is the general trend that is towards deskilling, so nobody should expect to find all the predicted symptoms of a deskilled labour force in particular firms at particular times. The same argument is employed in respect of the extent to which labour is able to influence changes. Short term influence will not be sustained (Armstrong, 1988 pp151-154).

Finally, Armstrong argues that many are able to refute Braverman's hypothesis on the basis of reskilling evidence, or the activities of labour, because they over simplify his argument. Braverman, really meant that the deskilling tendencies would eventually overcome temporary concessions such as those found by Jones (1982). Such a qualification, however, lays it open to the same kinds of criticism as Popper makes of Marxism. It is a 'reinforced dogmatism' since for Popper evidence cannot be accepted as falsification of the argument (see Popper, 1972 and 1966). The more specific consideration of empirical cases, however, may have identified other mechanisms by which jobs are defined and filled, which are alternatives to the limitations and claims of labour process kinds of analysis (See also Jones, 1988 p482n(2)).

The labour process perspective represents an alternative view of job design, arguing that the key managerial motive is that of control. This control is achieved by the deskilling of workers. The model involves a clear managerial strategy and rationale, raising questions for this book that must consider the relevance of the model to current approaches of work organization and job design.

Neo Fordism

Fordism proved itself to be an unpopular system with workers, witnessed by high turnover and absenteeism as well as large numbers of rejects of components and problems of stoppage and sabotage (Beynon, 1984; Pignon and Querzola, 1976). In 1982 Sabel predicted 'the end of Fordism' in a chapter of the same name in his book *Work and Politics*. The argument was one based on the notion that product demand was changing to more highly customized products from mass produced goods. The economies of scale possible under mass production have diminished rendering Fordism quite inappropriate.

Offering examples from Emilia Romagna in Italy, where artisans have begun to work in small workshops as interdependent units, Sabel suggests that there may be new trends toward the reorganization of work based on high trust (Fox, 1974 and 1985) facilitating a flexibility and adaptability in the workforce. Nevertheless, Sabel (1982) was aware that managers could exercise options of introducing flexibility by using innovative technologies, and at the same time, attempt to retain tight control through a detailed division of labour.

There may also be attempts towards job reform, such as job enlargement and job enrichment, but with no intention of increasing the levels of discretion of workers. For Fox (1985) this is an attempt '....to generate a high trust response from what employees experience as a low trust situation' (p110). Sabel refers to these kinds of approaches as neo Fordism, which tackles a key managerial problem; how to achieve motivation and yet not impose tight controls (Pignon and Querzola, 1976).

By maintaining seemingly high trust, but actually highly controlled, management working practices, the real agenda of neo Fordism is concealed. Here the objective of any job enrichment or job enlargement scheme is for purposes of the intensification of labour (Palloix, 1976). For others it is an adaption from the methods used in mass production consisting of modified systems of control that serve to intensify labour (Aglietta, 1987). It is also perceived by some as a means of reducing conflict in organizations (Ramsay, 1984).[6]

Labour process views and post Fordist models of work design are linked in their idea of managerial control. However, where Braverman (1974) claims that control is won by deskilling, neo Fordism offers a more subtle means of control in which work humanization schemes actually conceal the control methods. However, like the labour process approach, this still requires some form of control rationale and planning for successful implementation. One problem in locating the existence of neo Fordism in the organization of work is identifying which methods can be confirmed as concealing control motives, and which are implemented in the traditions of sociotechnical design. This should be apparent from managers rationales for new technological change.

The nature of current technological changes

The term 'technology' is used variously to describe forms of development over a period of years and at particular times in history. 'New technology' has referred to the latest developments in machines and production systems. New technology for the purposes of this project refers to microelectronic systems for the processing and communication of information and control. In this project three types have been selected for analysis. These fall into the categories of manufacturing technologies (specifically computer numerically controlled machines), newspaper printing technologies (photocomposition), and office systems, particularly systems for inventory control and accounting.

These three clearly are not the only sectors in which new technologies are employed nor are they the only new technologies available. Therefore it is important to see them in the context of general new technology developments.

One group of writers have identified four principal areas of new technology applications (Blackburn *et. al.*, 1985 pp14-20). The first area, robots, are potentially most useful in labour intensive processes such as packaging or assembly. Secondly, manufacturing process computation describes the use of computers to speed up processes, improve quality, and increase throughput levels in manufacturing. Office machinery is the third area, and is exemplified by word processing systems but includes a host of integrated computers dealing with the full range of clerical functions. The final area is electronics information handling which includes electronic mailing and funds transfer systems.

Microelectronic technology is by nature highly flexible and adaptable and has made changes possible in product markets. Mass production, economically viable at high volume throughput becomes increasingly less viable in markets requiring small numbers (batches) of variable products. This is due to the nature of the manufacturing systems used in mass production that would require constant resetting, while computer controlled systems are controlled by programmes that can be accessed and changed rapidly (Sabel, 1982). In the other areas of printing and office applications, the specific effect is different but the nature of the change is the same.

What all the technologies have in common is rapid transfer of information, which unlike the previous mechanical or manual processes provides, in theory at least, speed and flexibility. Such radical change in technologies ought to mean radical change for jobs, in many cases fewer people are required, and for those remaining profound changes in their jobs are likely.

However, it is difficult to generalize about the precise effect to jobs subjected to new technologies. The more general concept of 'automation' has been variously described as a device for saving mental labour, manual labour, and mental and manual labour simultaneously (Kaplinsky, 1984). It is perhaps a rather more subtle analysis to suggest that the present significance of automation, referring to current advanced technologies, is the presence of the control capability over jobs and processes (Bell, 1974).

Technological determinists argue, that once introduced, the new technology dictates work organization and job design. In this case the choice of systems can determine the organization of work. It may be possible to label certain technologies with certain styles of design. For example, Blauner (1974) argued assembly line technology dictates detailed work design, while more independently controlled techniques dictate more participative designs.

But the choice does not end with the choice of systems themselves, but actually extends to the way forms of technology are deployed. This is likely to create variation in job design and work organization, and the choice available in deployment may, therefore, result in political or social decisions. These decisions may be informed by managerial presuppositions involving issues such as work allocation and trade union response. The choice of new technology and the choice in deployment are therefore central questions in work reorganization programmes.

Political and social factors in new technology and automation

In the following quotations two perspectives are identified that are involved in the debate that surrounds technology and the socio-political dimension:

(i)
One sees technology as being unproblematic following certain 'natural paths' of development: the technology is neutral and bears no significant relationship to the social relations in which it was conceived or to work practices it involves (Kaplinsky, 1984 p133).

(ii)
The other view argues that technology must be seen in the context of its development, notably as an instrument where capital controls labour, and that it consequently carries with it four types of social relations. It involves hierarchical work patterns, is associated with increasingly fragmented tasks (the Babbage principle), it deskills the work, and it separates the conception of work from its execution (Kaplinsky, 1984 p133).

It is interesting that Kaplinsky accepts the relevance of a labour process view, seeing control over labour and the propensity to deskilling, as synonymous with the notion of a political dimension. The notion of technological determinism, which Kaplinsky describes as seeing technology as unproblematic, is a criticism that is levelled at the labour process approach, as well as other writers from different perspectives.[7] However, non technologically deterministic analyses are not the exclusive domain of the labour process perspective.

Indeed, some labour process analysts, for example Braverman (1974), may themselves be guilty of failing to consider, and oversimplifying, all the political factors that may be involved. To this extent Kaplinsky's work has limitations but is useful as a starting point.

Elsewhere (Wilkinson, 1983) it is argued that notions of efficiency and technical superiority have been used to cover up the real political agenda. Where such criteria are employed by managers they should be treated with suspicion:

....what our case studies show is that arguments about efficiency of new production technologies can serve as scientific glosses which conceal or obscure the political considerations which have gone into decisions on technical change and work organization (Wilkinson, 1983 p84: emphasis in original).[8]

Hence, while some theorists (Taylor, 1911) have identified 'scientific' methods of job design on the basis of efficiency criteria this underlies a political dimension. In Taylor's case this is supported by simplifying tasks to an extent that they could be easily understood by managers, which would mean important increases in control. Although the method suggests, that because of its scientific neutrality, it has freedom from values, the reality is quite different.[9]

Wilkinson's implicit use of an action approach explores the functionalist assumptions of technological determinism. He reveals that managers and engineers do not always assume the role of neutral intermediaries that many assign to them, but become '....creative mediators between potential and actual technologies....' (Wilkinson, 1983 p19), and preside over decisions that will have far reaching effects in social, political, technical, and economic terms. However, it is suggested that this use of an efficiency argument can be positive when it recognizes the need to achieve the cooperation of the workforce, after all this too can be justified in efficiency terms:

> Managers use of the efficiency argument is thus not <u>always</u> sinister and consciously understood. Rather, it seems that in part, managers feel most secure when all their actions can be justified in terms of increased performance (Wilkinson, 1983 p86: emphasis in original).

Therefore, considerable care needs to be taken not to assume that the only considerations, when it comes to the deployment of either technology or personnel, are those of economic and technical significance. Decisions are also based on a range of political and social criteria. In studies of job definition it is therefore essential to take into account the decisions made in choosing new technology and then deploying the technology. Political motives at these stages may have direct implications for the way jobs are designed and work is allocated.

The distribution of skill and technological change

It has been shown that the introduction of new technology will effect jobs, and like the new technology itself, there is choice in the way work is to be organized. Job definition decisions may be based on the form of technology, the way this technology is deployed and the favoured division of labour. This in turn involves the distribution of skills. The issues that arise here are possibilities of repackaging skills, redistributing skills, and the types of skill that survive technological change, or develop as a result of technological change.

The term 'skilled' has been referred to as '....an objective characteristic of work routine and job knowledge....' (Littler, 1982 p7). While others have classified skill on the basis of whether the work is 'specialist' or 'specialized' (Friedmann, 1961 pp84-88). Thus providing two very broad categories of a professional group of people, and of workers involved in detailed tasks. The problem is clearly not just one of lack of specificity of the groups, but it also lies in the lack of coverage and the assumption that all jobs can fall into these categories. This is a difficult model to apply, since it is quite feasible to have jobs that mix tasks and discretion levels. It cannot be assumed that if a job has a specialized organization, that it has no specialist aspects. Of course, the reverse also applies.

However, the main problems of such definitions of skill lay in the absence of social and political considerations. Skill cannot be only an objective product of the tasks performed in a job. It must also be a product of elements of social construction in the establishment of which tasks are skilled, followed by decisions about how these tasks are incorporated into a job. Some approaches see this definition as favouring men over women (Phillips and Taylor, 1980). Hence, skill becomes synonymous with control.

Traditional craft occupations were distinct from other jobs because they embodied a level of autonomy in the work, allowing the worker to make personal decisions about the conduct of his/her work:

> A relatively skilled position was one of trust, where the worker was granted a sphere of competence within which decisions, whether routine or complex, could be taken by the worker himself. This 'guaranteed autonomy' is the essence of the traditional craft occupation within which the workers themselves control the productive process.....Psychologically it is an encouragement to self direction....It is social not technical. The centre of the technique is not complexity, but autonomy and freedom (Blackburn and Mann, 1979 p292).

This and other similar viewpoints make the vital link between skill and control, the autonomy of craftsmen is an example which is often referred to. However, work need not have such a high status to contain skill, and it is possible to suggest all jobs have some skill content. Even work that is designated as unskilled work may have elements of 'tacit skill', that is elements of skill specific to a job but not tangible nor recognized by managers as skill (Kusterer, 1978; Mainwaring and Wood, 1984; Libetta, 1988). A range of external factors play important roles in deciding on classification of a job as skilled or unskilled. These include skill in the person, skill in the job, and a political definition of 'skilled' which refers to the effectiveness of a group of workers to defend their work as skilled (Cockburn, 1983).

The questions of the technical input itself, and the fusion of conception with the execution of tasks, cannot be seen in isolation as the sole bringers of legitimate skill classifications. We have to be prepared to accept much broader and varying ideas about skill, and be alert to the multitude of ways firms may classify skilled work and whatever other categories they may have. Rainbird (1988) suggests that for trade unions, the recognition of new skills, and definition and defence of jobs as such, is essential given the trend towards the merging of jobs as traditional demarcations are broken down.

Within the new technology and skills debate, there are two contrasting and recurrent predictions of the skill consequences of further automation. Braverman (1974) and his school predict that the workforce will become deskilled and divided under the advance of technology. The opposite interpretation is that new technology brings in its wake, more involving and rewarding work (Hirschhorn, 1984). Both these views have been criticized for failure to consider factors that are necessarily involved in the process of introducing technology. Nevertheless, Davis and Taylor (1979a, pp108-109) have also argued along similar lines, suggesting that technology has created a new demand for skills of mental and perceptual ability, when previously it was manual coordination that was required.

One useful aspect of Blauner's (1964 pp169-171) analysis is that it provides a contrast between different stages of the advancement of technology. On the one hand he sees assembly line technology as highly restrictive of a workers creativity. On the other hand, his empirical studies of process technology in a chemical plant, suggested to him that control is increased in this type of environment due to the nature of the work being performed. It is strongly reminiscent of the kind of supervisory function that Touraine suggested will become the norm with advanced technology systems (Touraine, 1962 in Littler, 1982).

The opposing view, held by Marxists in the labour process tradition, is that of the organization of the division of labour as a capitalist imperative, inevitably requiring constant rationalization as technology advances. This means that labour is employed in increasingly specialized roles, based on the ever increasing separation of the workers conception (or creativity), from the simple execution of highly divided work. The result is alienation[10] and lack of control.

Kaplinsky (1984) concurs, seeing the evidence as demonstrating that the technology is employed because of a motive of control, and new advances create still more opportunities for removing craft control and for deskilling work. In doing this he uses Wilkinson's (1983) argument, that efficiency reasons really cover up political aims. However, he does this only in respect of deskilling, and fails to acknowledge the possibility that enhancement of work may be packaged with similar motives, but conceals these behind a facade of apparently beneficial changes. Kaplinsky does not completely deny the existence of new skills, but views them with considerable scepticism. This is implicit in his account, but he is quite explicit when he explains that new skills are unlikely to be great enough in numbers to replace displaced skills. Even if new skills are evident they too are subject to the same labour process:

Although there is no doubt that over the long run there is an increase in the 'average' skills of the labour force, the emerging skill requirements of automation technology thus suggest an increasing polarization between the very much higher skills of the systems designers, product designers and management, and the degrading and deskilling of traditional skills (Kaplinsky, 1984 p137).

These perspectives overlook the possibility that there may be technological reasons for retaining traditional skills. However, others have recognized that where the technology fails and cannot automatically right itself, workers have to put it right. To do this traditional workers need to be endowed with the appropriate skills involving both larger jobs and increasing responsibility:

If as I believe to be the case, error is inevitable in automatic systems - if there are always to be modes of failure that cannot be automatically regulated by feedback-based controls - then learning must be instituted in order to prepare workers for intervening in moments of unexpected systems failure. Failure in turn is a specific example of discontinuity and developmental change. Thus we define postindustrial work as management on the boundaries of systems and physical realities. Historically, we could see the worker moving from being the controlling element in the production process, to operating the controls, to controlling the controls (Hirschhorn, 1984 p73).

Hirschhorn's ideas are close to those of Alain Touraine, who argues that there are three stages of capitalism; the first refers to the old system of craftwork, the second involves the development stage of mechanization and is actually a period of transition, in the third and final stage the development of mechanization is complete, and the workers role becomes a supervisory one in which the machines are maintained and controlled (Littler, 1982 p13).

Another perspective acknowledges that deskilling has occurred but social processes have prevented proletarianization of the workforce (See Littler, 1982 p14). Clearly this may include methods of preserving skilled status through job enrichment and job enlargement. The notion that such methods have been adopted to shield capitalism from an increasingly deskilled (and therefore increasingly polarized) workforce can be found in both Braverman (1974) and Zimbalist (1975).

Edwards (1978 p109) suggests a far more complex system of simultaneous deskilling and reskilling of the labour force. Labour process analysts tend to give priority to labour saving above capital saving, and, while this may be driven by pressures of class struggle, this is still only one pressure amongst many others. Possibly more important imperatives to consider for the capitalist are competitive pressures; and in this situation whether savings via technological innovation come from labour or capital may matter little. We then arrive at a position which Littler elaborates as follows:

Recognizing capital saving innovations means recognizing that the development of technology and science has resulted in increased work complexity as well as the lodging of skills in machines. This means that the influences on skill can operate in two opposing directions and that the outcome is problematic (Littler, 1982 p14).

New technologies can be quite different in nature and some are capable of removing routine work and highlighting skills, on the other hand they may substitute the need for skill or pass the skill from the shopfloor leaving the shopfloor with fewer skills and more mundane work. However, it is also important to bear in mind that the skill equation is even more complex. Much of the work that sees technology as potentially enlarging jobs does not necessarily make distinctions between the skills that are added.

If tasks are added to jobs at the same level as a workers' existing skills then the main implications are the quantity of jobs remaining and the choice of workers for those jobs. However, if the job task content changes involving a 'vertical loading' of tasks, that is, an addition of higher level skills in planning, design or supervision, then the implications are not only the allocation and job loss issues but go wider challenging the existing hierarchy and command structure.

For these reasons skill outcomes may be extremely varied, dependent on the technology that is chosen, the way that it is deployed, and the way that tasks are made into jobs. There are also important other elements of influence involved which go beyond managerial decision making, such as the influence of the workers themselves and of trade unions (Friedman, 1977).

New technology and industrial relations

Managerial strategies and new technology

Much has already been said about management responses to technology. There are a range of possible approaches and outcomes according to managerial preference and motivation. For Braverman one management method, that of Taylorism, prevails over all others and the use of technology is based on a deskilling of the workforce and achieving greater control. Buchanan (1986) is one of many who challenge this notion claiming:

> The capabilities of new technologies are not determining but underline characteristics, which open up new opportunities for producing services, processes and opportunities of work organization. These characteristics may also create new forms of constraint on managerial action (p69: emphasis in original).

The technology criteria that management are working with are laden with choice within the parameters of the capitalist imperative. The use of technologies can be varied and do not necessarily have to be one track as Braverman suggests.

There is however, another possibility that emerges that there may not be carefully considered reasons for the introduction and use of technology. In the companies studied by Burnes (1989) there appeared to be little planning of a financial, technical or job nature. Similar suggestions are made by other authors (Rose and Jones, 1984; Hyman, 1988; Jones, 1988). Counter to this

Child's (1985) account suggests there are strategic processes and these can be classified into four types of managerial initiative. These are: the elimination of direct labour, the spread of contracting, the dissolution of job or skill demarcations and the degradation of jobs through deskilling. Child sees these as managerial objectives in changing the labour process but unlike Braverman (1974) does not claim that the primary reasons for the use of new technology is to achieve these objectives. Objectives such as those identified may be present in complex combinations and may hide behind aims of improving competitiveness.

The strategy a company chooses is informed and influenced by factors such as government policy, institutional culture, product and labour market conditions. Even if a managerial strategy exists it is unlikely to be straight forward. Therefore, strategies for new technology changes are influenced and informed by a range of externalities that may change the imperatives and the direction of policy (Salaman, 1986). The more considered the approach the less likely that it will drift from the declared policy. Hence a 'partial strategic approach', one that involves some aspects of planning but excludes job design and work organization issues is likely to be carried off course by the myriad of influences prevailing upon it to a greater degree than an approach that considers all aspects.

The fluid nature of strategic planning suggests that there are potentially considerable possibilities for trade union influence.

Workers, trade unions and new technology

Managers have an interest in involving trade unions and workers in discussions involving cases of technological change. This is because trade union agreement is important for the smooth implementation of schemes, but trade unions may also raise practical issues providing a useful source of information and guidance for managers. However, not all managers view trade unions as a useful resource in this way. Some see their involvement as an infringement of managerial prerogative, and as such, involvement is something to be minimized. In the latter cases trade unions have to form a strong case for their inclusion in new technology discussions. This section considers the case of trade union involvement and the effectiveness of trade unions in negotiations related to new technology introduction.

F. W. Taylor, the instigator of an allegedly scientifically neutral system of job design, mistrusted trade unions. Scientific management was to render labour unions and strikes redundant because the system would apply to individuals not just groups of workers (Nadworny, 1955; Kelly, 1982). There was a belief that the incentive system would place the question of earnings at the command of individual workers so that they would cease to rely on trade unions for support in achieving pay increases. Henry Ford shared similar views and was prepared to pay large bonuses

to avoid unionization. In Taylor's case, many of his followers
were opposed to this exclusion of unions and after Taylor's death
chose to accommodate them. Nevertheless, historically the fear
of trade unions was enough for it to become an important issue
in these highly influential systems.

Despite the very real concerns of the systems exponents, others
tend to play down the role of unions in Taylorism (Braverman,
1974). Braverman's problem then is his tendency to assume a
constant sum of power in which managers are omnipotent leaving
the trade unions and workers impotent. The unswerving belief in
Taylorist philosophy precludes the possibility of there being
meaningful worker resistance (Friedman, 1977).[11] Several factors
may influence the effectiveness of worker resistance which
including plant size, organization, type of work and the
homogeneity of workers, both vertically (in terms of levels of
skill) and horizontally (in terms of age, sex and race). But
Friedman (1977) argues that trade union resistance to 'direct
control strategies' have resulted in 'responsible autonomy' or
more participative methods of work organization.

Reports of trade union responses are varied. Wilkinson's (1983)
case studies found trade unions devoid of strategy and 'worker
initiatives remained largely at a subterranean or individual
level', but this had proved a formidable force:

Our case studies repeatedly made it clear that the strength of subterranean challenge
to management designs should never be underestimated, but nonetheless one might expect
trade unions to take a more serious interest (Wilkinson, 1983 p98).

While Wilkinson concludes suggesting that the inadequate
bargaining structures need revising, other writers have disagreed
on the issue of union influence suggesting that this can be
significant and varied (Clark, 1989; Price, 1988).

Work by Price (1988) provides evidence that there are two broad
groups comprising cooperative relations and conflictual
relations. These were represented in his study as eight companies
falling into the former category and four in the latter. The
former, marked by regular consultation, demonstrated that the
unions could achieve their objectives and be expected to be
involved in issues of plant layout and even allocations of jobs.
Price even records that APEX at one company were able to draw up
a new technology agreement centred on issues of health and safety
and job design, although what is meant by job design is not
discussed. In the companies displaying conflictual relations
consultation was rare, levels of trust were low and there was
considerable contempt for the union. Managers would accuse shop
stewards of not relaying the information to the workforce and
shop stewards replied suggesting the information simply was not
available from managers. There are examples of the denial of a
signed new technology agreement and of redundancy without
consultation.

Price's work suggests that there are dual possibilities in the response of unions and it is inadequate to suggest all trade unions are ineffectual at plant level. However, in all the companies enjoying cooperative relations, what was not discussed informs us of the considerable limitations to union influence. For example, the choice of new technology was never an issue, this was clearly an area in which managers wished to maintain an exclusive interest and often unions saw as being outside their interest. Beyond this, all consultation was management led and took place within a framework designated by management. In general the issues the trade union were able to influence were health and safety and number and nature of redundancies (voluntary or compulsory). Job allocation and job design are rarely discussed although clearly there are some important exceptions.

It may be that the TUC guidelines on new technology *Employment and Technology* (1979) are too ambitious to be realistically achieved by plant level branches but it may also be equally true that union representatives are more than happy with a few minor concessions that management offer them. Davies (1984) documents a case where the trade union was 'thoroughly satisfied' with their involvement in discussions. However, consultation did not take place until four years after the initial management decision to invest.

The kinds of rationale managers put forward in these situations were that unions did not have the knowledge or expertise to comment, they may include destructive elements, and the preference was often to leave out unions since employees were more manipulable outside trade union organization. There was also the view that unions were weak and there was little point in including them on this basis. However, many of the unions in Davies's (1984) study expressed their commitment to new technology and more efficient methods. Given this positive attitude it seems that management had nothing to lose and probably significant gains to make by speaking to the trade unions (Jones, 1988; Rainbird, 1988; Batstone *et. al.*, 1987; TUC, 1979; Northcott *et. al.*, 1985).

The emphasis of the TUC's report *Employment and Technology* (1979) was that new technology offered considerable potential benefits to workers in offering opportunities for improving their quality of working life. The role of trade unions was to see that the benefits were distributed equally between workers and employers. One way in which trade unions have sought to achieve this is by the use of new technology agreements (NTA). However, the success of most unions in achieving these is limited and rare, the scope is equally limited (Jones, 1988). The involvement of trade unions at design stages and in system planning '....is almost unheard of....' (Jones, 1988 p473). Even in cases where trade unions are strong, and the union has a reputation of supporting the business, efforts for inclusion in decision making are rarely realized (Batstone *et. al.*, 1987). Williams and Steward's (1985) work on NTAs reveal few agreements of this nature in the UK. Those that did exist were limited in scope and often contained a clause that exempted categories of information on the grounds of confidentiality. NTAs tended to exclude key areas such as the quality of working life and job design issues whilst the unions weak bargaining position was an important factor in the low level of acceptance by managers.

The possibilities for trade unions to influence managerial decision making in respect of new technological changes is not in question. What is questioned are the abilities of different trade unions to put together an effective case for inclusion in discussions of new technology. Secondly, for trade unions to consider issues outside the quantitative areas of pay levels, terms for redundancy and redeployment and the numbers of jobs to be lost. Although there is evidence of trade union involvement in the former issues, little evidence exists of union involvement in the decisions of which system to buy and of how to organize work and design jobs. New and existing case studies should assist in answering the questions here.

Flexibility, new technology and job design

The issue of flexibility is important because it pulls together the issues of skill, flexible technologies and the choice of deployment of staff. But it also suggests a possible alternative means of job design as a response to new technology. Some writers emphasize the reorganization of the firm towards more flexible operation. Part of the process of the redesign of jobs may involve a wish to increase the flexibility of the workers. If there is a connection between redesigning jobs for flexibility and new technology, this is of fundamental importance since the flexibility issue may be influencing and informing managers in their approach to the repackaging of jobs.

The flexibility debate necessarily refers not to labour alone but also organizational and external factors. Preece (1987) suggests that there are four areas of organization that influence flexibility: labour - the division of labour between skilled and unskilled work, production - the systems employed according to products required, technological flexibility - the available technologies offering flexibility, and structural flexibility - the extent to which an organization displays a formalized or bureaucratized structure as against an adaptable structure. In Fordist organized plants the labour is divided up into its component parts, the systems of mass production require little production flexibility and the technology is geared to producing the same product *en masse*. The structure of the organization is highly formalized and bureaucratically organized around rapid and closely controlled, machine paced, production. The system has an inbuilt inflexibility that appears appropriate for mass production (but for the considerable labour problems experienced). However, the inappropriateness of the system is apparent not only from labour opposition to demeaning work (Pignon and Querzola, 1976) but is also driven by a shift in product markets to a demand for more variable products (Sabel, 1982). New technologies may be able to facilitate the flexibility required in technical terms of accuracy, adaptability in programming and reliability (Preece, 1987) but others argue that this requires a flexibility in labour or a 'functional flexibility' (Atkinson, 1985).[12]

However, the use of multiple labour markets is a concept that several writers have identified (for example, Edwards, 1981; Piore, 1980; Doeringer and Piore, 1971; Atkinson, 1985). For the most part these accounts offer variations on a theme. Broadly there are two or more labour markets operating, providing for workers who are required as of key importance in the organization and on the other hand, a market providing workers purely for their labour power who are largely an untrained group.

Technology may offer opportunities for work to be simplified as well as made more complex. The resulting scenario may then follow a dual labour market situation. For example, combinations of technologies may simplify some work whilst creating more complex jobs elsewhere. Given that this is the case employees can be slotted into primary and secondary groupings. The primary workers as the valued employees who are not easy to find in a labour market will be given good employment terms. Employees regarded as secondary become dispensable, and are easy to top up or reduce since the nature of their work is very simple. Although it should be noted that challenges to the straight forward replacement of employees are brought by a number of writers (Mainwaring and Wood, 1985; Libetta, 1988; Polyani, 1978; Kusterer, 1978) who present evidence to suggest that a large part of a jobs content is tacit, that is, unarticulated worker knowledge learnt on the job. The removal of one unskilled operator for replacement by another from outside, or the creation of a new unskilled job may cause teething problems and be costly in error while the new worker learns those hidden elements of the job that become important for the smooth operation of the work (Rainbird, 1988).

This form of flexibility in labour means something more that the expansion of responsibilities within an individual's job to meet the demands of the system. It also means demarcation breaking and the merging of jobs and this has become an increasing tendency (Jones, 1988; Rainbird, 1988). Blurring of the distinctions between direct and indirect labour usually resulting in a reduction of direct labour is also commonly noted (Rainbird, 1988; Coriat, 1987; Arthurs, 1985). The process may involve multi-skilling of craftsmen such as training mechanical fitters in electrical trades and electrical workers in mechanical trades, or adding electronic knowledge to the responsibilities of the electricians' job. It may also mean taking the operators jobs and adding some maintenance tasks (Jones, 1988). Hence, while functional flexibility refers, in a broad sense, to the acquisition and application of a broader repertoire of task components to a job, 'flexible specialization' (Piore and Sabel, 1984) refers the acquisition of specialized, more highly skilled, components which may include aspects of supervision, planning and design. In other words, flexibility in the former is a loading of tasks along the horizontal axis of equal skill, the latter is an addition along the vertical axis of higher skill and more specialist work. However, it should be noted that defining flexible specialization has proved problematic (Jones, 1990; Wood, 1989).

In any situation where the task content of jobs becomes broader there is likely to be a merging of functions within jobs that necessitate the breaching of demarcations. Here conflicts between different trade unions are likely due to one union attempting to defend the established demarcations and another recognizing the benefits for its members to take on these skills. One solution is union amalgamation (Rainbird, 1988) but it is unlikely that this will solve the problems since conflicts are also likely within unions that represent different skill levels. Unions such as the AEU may find themselves in a dilemma since semi-skilled workers may begin to take on skilled maintenance work and internal divisions that develop may be more damaging than inter union conflicts.

At another level the threat is to supervisory staff who lose their managerial control function through the flexibility of labour and of new technology. On the one hand, through 'responsible autonomy' (Friedman, 1977) initiatives, workers' jobs may be 'loaded vertically' including elements of self supervision such as quality control. But more importantly more elaborate systems enable closer surveillance of workers through the machinery (Jones and Rose, 1986). Jones (1988) suggests this involves supervisory staff taking on more of a technical or advisory role but here several studies have shown that supervisors lack the training and expertise to provide sound technical backup in areas such as computer numerically controlled machine tools (Burnes, 1989; Batstone *et. al.*, 1987).

Increases in secondary or numerically flexible workers are becoming more common particularly with the shift to more part time, casual, and short term contract work. This kind of work is often associated with particular social groups such as women, racial minorities, migrant workers and the low skilled.[13] However, the role of trade unions can be important here although what is worrying for unions is that the tendency to merge work roles will result in the division of trade unions and render them incapable of effective opposition (Child, 1984). On the other hand the retention of demarcations in some areas and the deskilling of work elsewhere may mean a division that is orchestrated and maintained as a means of protecting existing workers artificially.

Summary and conclusions

In this chapter the theoretical issues related to job definition and technological change have been assessed to provide a framework for consideration of case study material.

A useful starting point is to consider the technological context. New technology has meant that changes to work organization have been necessary. The nature of the change has been based on the much greater flexibility of microelectronic technologies. The rapid transfer of information has enabled products to be more varied unlike conventional technologies in mass production which were capable of only producing large quantities of goods in mass production cycles (Sabel, 1982). Attempts have been made to show that the advance in technology has also meant an increase in worker autonomy (Blauner, 1964). This represents a technologically deterministic view and other theorists have attempted to show that there is more choice available in new technology. This choice not only occurs at the level of the technology but also at the level of work organization once the technology is chosen.

If new technology offers choice in the way work is organized what options are there for designing work? The ideas presented in neo classical job design theory (Taylor, 1947) are adopted by writers of the labour Process interpretation, notably Braverman (1974). Taylorism involves a simplifying of work so that each job is comprised of repetitive tasks which reduces worker autonomy to a minimum. Braverman highlights this simplification of work claiming that this is a means by which managers achieve control over workers. Neo Fordist writers follow in the steps of the labour process in the sense that both accounts are economically deterministic and each see managerial control as the key motive. However, neo Fordist ideas (Palloix, 1976; Aglietta, 1979) deviate from those of the traditional labour process view in the methods they see as being employed for control. The latter highlight deskilling while the former suggest there are far more subtle means which may involve job enlargement or job rotation, but the managerial intention of control remains the same.

However, each of these accounts is deterministic, an alternative approach is required which allows consideration of more variable socio-political criteria. The sociotechnical approach (Davis and Taylor, 1976) is such an interpretation in which in which the humanization is genuine but not independent of, nor dependent on technical and economic requirements. However, each of these are strategic approaches planned by managers to produce particular outcomes and these can only be applicable if planning is a part of the job definition process.

It is important to eastablish to what extent managers are pursuing strategic approaches.

It is implied that job changes may be strategically planned or job considerations may be dealt with in an *ad hoc* manner. Planning itself may be dependent on the way managers view new technology, some believing that the technology dictates work definitions, others acknowledging the existence of choice. Indeed, a range of approaches suggest that there is considerable variance in the skill levels that emerge as a result of new technology and this would seem to demonstrate that there is choice in the way work can be organized. Consideration of skills does not tell us much about planning but is an appropriate way of considering claims that new technology will result in more highly skilled work (Hirschhorn, 1984) or conversely deskilled work (Braverman, 1984). A third possibility is that a mixture of job definitions occurs (Edwards, 1978). It is possible to consider these various analyses in each of the case studies.

The addition of skills to jobs may mean the merging of occupations such that tasks are removed from one job and added to another. Since this implies an advantage for one worker and disadvantages for another there is a possibility of conflicts developing. This is likely to be a further problem in job definition decisions and it is useful to consider the extent to which this has occurred, resulting in trade union disputes with each other. A further possibility is that new technology may introduce greater flexibility into organizations suggesting that workers themselves need to be more flexible. Functional flexibility or flexibility across a range of tasks (Atkinson, 1985) means a further option is available so that managers can define work broadly, using workers according to peaks and troughs of business activity.

Given the existence of choice it is appropriate to consider not just the extent to which managerial strategies are employed, but also to consider for new technology change processes, who else is capable of influencing and diverting that change. In the literature, trade unions are seen as a group that may be able to influence the course of change (Clark, 1989; Price, 1988). Some writers question the effectiveness of trade unions (Wilkinson, 1983), while others agree with this lack of effectiveness but point to efforts that have been made, but have failed due to management intransigence (Williams and Steward, 1985). This is a further aspect that needs to be taken into account in case study analysis.

In the next chapter existing case studies are considered in an attempt to find empirical data that provide answers to the questions raised in this chapter. Hypotheses are generated according to the theoretical questions here and the adequacy of the available data.

Notes

1. Since, Taylorism is documented in a range of works (see for example Warr and Wall, 1975; Littler, 1983 and 1985; Littler and Salaman, 1984 and 1985; Kelly, 1982; Rose, 1978 and 1988; Nelson,1980; Nadworny, 1955; also Taylor himself 1947) where Taylor's work study techniques and managerial philosophy tend to be the recurring themes, only a brief resume of the relevant issues is required.
   Others contributing to the debate are, for example: Whitaker (1979); Thompson (1913); Cadbury (1914); Maier (1970).

2. In Fordism the control mechanisms were largely contained within the machinery rather than imposed by supervisors as in Taylorism.
   Fordism emerged as a new model popularized by Henry Ford. The difference between Taylorism and Fordism being that Taylorism, as practised, was concerned only with the management of labour, while Fordism dealt with the whole production process (Maier, 1970). Similarities between the systems are recognizable since they are based on similar principles, but they are fundamentally different in the way they are applied. For example, Ford's Highland Park plant had a time study department and Fordism had clear divisions of labour but Henry Ford denied the use of any established management system:

   > The Ford approach was to eliminate labour by machinery, not, as the Taylorites customarily did, to take a given production process and improve the efficiency of the workers through time and motion study and a differential piece rate system of payment (or some such work incentive). Taylor took production hardware as a given and sought revisions in labour processes and the organization of work; Ford engineers mechanized work processes and found workers to feed and tend their machines (Hounshell, 1984 p252).

3. Such 'piece work' systems were known not to work because of bad management practice which took the form of 'rate cutting.' This describes managers realizing a job can be done quicker and progressively cutting the rates paid for the work. A certain way of encouraging systematic soldiering, not eliminating it. For Taylor this represented an abuse of his system and he complained that there was a failure to implement his system in full.

4.
   > The effect of automation and increasing use of technical aids can only be determined on the basis of an analysis of its interaction with organizational and labour factors and it has come to be looked at in view of an overall strategy in the plant or company (Sorge, 1983 p50)

   See also Cockburn, 1983 who argues that there was significant labour influence in the printing industry.

5. See also other critical papers in Wood, S (ed.) (1982) *The Degradation of Work?*

6. Pignon and Querzola (1976) present examples of where there are apparent possibilities for worker control. The illusion is achieved by relaxing supervisory authority and encouraging suggestions from workers; pay and bonuses are related closely to production levels. In this way worker initiative is harnessed and control maintained. In another company the authors find a decentralizing of authority and job expansion which incorporates responsibilities in designated areas which make it easy to locate the person making the mistakes. Hence positive incentives are given in the form of job enlargement but negative incentives are built in to maintain close control.
    What is achieved here is more of a neo Taylorism; but what neo Taylorism and neo Fordism have in common are the ability to artificially reduce conflict by involvement which in turn gives access to worker initiatives, but alongside this subtle but effective methods, such as negative and positive sanctions, keep firm control of the work.
    See also Wood (1989) on the diversity of definitions of neo Fordism.
7. The criticism is applicable to all those who apply a unitary model - for example either one that argues technology is introduced to set a particular labour process or one that sees a relationship between the technical demands of a firm and the technology chosen. The former acknowledges a political dimension but tends to retain this as the exclusive purview of managers - the technology is unproblematic at a slightly later stage than the latter, in the latter politics are simply not involved.
8. Compare also Pignon and Querzola (1976).
9. See also Collins (1985), Chalmers (1982) and Mulkay (1979) on the questionable neutrality and value freedom of science.
10. Alienation is the term used by Marx to describe the denial of self actualization through work which causes estrangement from work.
11. Zimbalist (1979a) suggests that Braverman was aware of the significance of worker involvement but felt that worker resistance would mean the hastening of technological change and so the decision was taken not to get involved.
12. See also Piore (1982) and Piore and Sabel (1984) on changes in product markets.
13. See Gordon *et. al.*, (1982) on racial and gender segregation, Piore (1980) and Kamata (1982) on the migrant worker and Walby (1989) on the use of women as a flexible source of labour.

3 Existing case study analysis

Introduction

Chapter two dealt mainly with theoretical issues in the literature concerning work organization, new technology and industrial relations. This chapter examines existing case studies to discuss further these issues, but with the added consideration of empirical case study material. Incomplete or missing answers lead to the hypotheses which the fieldwork tested in new case studies of six companies.

In this chapter, case study analysis focuses on three issues related to job definition and new technology, and represent the key areas of interest considered in chapter two. The intentions of managers, from the point of view of their rationales and implementation of policy, are examined to establish directions of change. Secondly, the outcomes in the design of jobs are considered to examine the case for generalization of trends, toward reductions or increases in the skill content of jobs. Here the allocation of workers to jobs is also taken into account. Finally, the extent of trade union involvement in work organization is dealt with. The key questions here are the depth to which trade unions are involved, particularly in job design issues, and the potential and actual conflicts that may emerge where demarcations are breached due to a redefinition of jobs.

The hypotheses presented at the end of this chapter are tested in the following chapters (five to seven) in the staffing and use of three main technologies: computer numerically controlled machines, photocomposition technology in provincial newspapers and accounts and inventory systems in office environments. These and accounts and inventory systems in office environments. These groups have been followed here, as far as is possible, for ease of comparison.

Case studies in metalworking

In the metalworking case studies computer numerically controlled machine tools are singled out for analysis following other work in this area (Jones, 1982; Noble, 1978; Batstone *et. al.*, 1987; Braverman, 1974; Sorge *et. al.*, 1983). However, the complexity of this form of new technology requires description.

The forerunner of the computer numerically controlled machine (CNC) was the numerically controlled machine (NC). Early machines had tape control mechanisms consisting of a magnetic or punched tape which was fed into the machine and contained the information necessary to carry out an operation, a new tape was entered for the next operation. Considerable advantages were realized through NC since seventy per cent of a component's time is spent in the machine as opposed to forty percent on conventional machines. Thus saving the remainder of the time spent, when the component is either moving from one machine to another, or simply waiting in a queue for a machine, known as 'floor to floor time' (Thyer, 1988 p11).

CNC is faster because a whole program to machine a component can be entered and stored in the memory of a machine's computer (rather than dealing with one operation at a time). The control unit receives signals advising the previous operation is complete before feeding the information for the next operation. Most CNC machines embody manual override facilities which are useful if there are problems with a piece of work or program. In some machines there is a simultaneous graphical representation of the component on a visual display screen contained within the actual control unit. Direct numerical control (DNC) is a more advanced form of computer control in which a series of CNC machines are all linked up to a central computer which then delivers information to each machine through 'on line' facilities.

CNC machines are particularly suitable for the manufacture of small numbers of components (batches) requiring a wide range of operations. Their advantages are: faster production rates, quicker set up times, reduced lead times, greater cutting efficiency, better quality, less inspection, less handling and ease of design change (Thyer, 1988 Chap. 2).

In terms of the work that is required of the operators there are several stages which can be seen as distinct tasks. Programming for the production of parts is often seen as the first stage,[1] but even before this the specification of the part has to be prepared before it is translated into computer language. After the program is prepared, the machine then has to be set with correct tooling and attachments. Many machining centres now have tool carousels capable of selecting the correct tool automatically. The program is then 'proved out' or tested. Amendments may be made, during the machining process, to the machine and the program. Prior to the machining process the blanks have to be loaded and after the process is complete the machined parts are then unloaded.

CNC machines not only contrast with conventional machinery in the advanced nature of the technology, but also in the make up of tasks and skills involved. There is potential for down grading skill on the shopfloor with CNC machinery but this need not be the case. In conventional operation, considerable knowledge is required to handle the production of a component from start to finish. Knowledge is also required to deal with the range of tasks involved in CNC. However, there are a number of choices in the division of labour. Jobs may be packaged to include a full range of tasks or small groups of tasks, and it is these decisions that are central in the management of new technological change.

Managerial intentions

This section explores the strategies and rationales of managers in existing metalworking case studies. One question centres on the dichotomy between labour process and other writers, on the presence of strategies with underlying control motives. Other writers have suggested that any formalized strategy is absent in the introduction of new technology. In these cases rationales are often poorly defined but reflect ideas about the capability of new technology to improve competitiveness, efficiency and effectiveness.

Although, Braverman (1974) cites numerical control of machine tools as a principal illustration of how deskilling has taken place, his account is lacking in empirical detail. However, other writers in a similar vein, such as Shaiken (1985) and Noble (1979), do provide evidence to support Braverman's claims. These writers follow Braverman by claiming that the labour process created within a NC machining area has avoided the use of skilled operatives developing new skills to program the machines. Instead the labour process is broken down into its component parts, separating conception from execution.

The significance of this form of organization for Braverman's theory is that the process is engineered to promote managerial control in a Taylorist (or perhaps Fordist) vein. After the planning and programming functions have been completed what is left for the operator is a job requiring little skill. The worker becomes '....a monitor rather than a participant' (Shaiken, 1985 p67). Similarly, Noble (1979) suggests that the development of NC machinery was based on managements wishes to control and reduce levels of worker involvement in the production process and therefore keep a tight rein on the workforce; even where the consequences might be to operate at a less than optimal efficiency.

Braverman, Shaiken and Noble all make the general point that managerial intentions play a key role in technology introduction, but they are all reporting on the American experience.

In the UK more varied accounts can be found. Wilkinson's (1983 p55ff) study of a machine tool manufacturer discovered that the majority of the CNC machines (eight out of nine), were designed with the intention that programming should be carried out away from the shopfloor. One CNC machine was capable of being programmed as well as having a facility for proving via a manual data input (MDI) facility. The procedure in choosing the technology was to buy the machine and then find someone to operate it, going outside the firm if the necessary skills were not available inside. The decision to buy the machinery was based on technical reports prepared by production engineers which were then put before senior management to make the final decision. Wilkinson highlighted the political dimension in production engineers' proposals. One engineer extolled the advantages of operator programming in CNC, another presented the potential use of the machinery as a means of separating programming and operation.

In terms of job design, management apparently had not planned the allocation of tasks, preferring to avoid any commitments and to keep their options open. Those involved in the decision making function were the managing director, manufacturing director and personnel manager.

Similarly, Jones (1982) contradicts the Braverman notion of strategic planning for deskilling and control.[2] Neither were labour costs a motive of CNC introduction in his study, but most commonly cited were reductions in machinery and 'floor to floor' time, quality, repeatability and the scarcity of highly skilled labour. Similar motives were identified by Burnes (1989) in his study, although one firm out of the nine studied by Burnes did wish to win greater control over the workforce. Neither study found much evidence of attention to financial and investment planning.

In the latter study it is suggested that often a main concern of management was to make changes conducive with the existing organizational structure which often meant rigid demarcations in working practices. Decisions were also motivated by the idea that CNC was easy to operate,[3] but this underestimated the skill requirements, and failed to appreciate problems associated with coordinating CNC with other traditional areas. The outcomes of Wilson and Buchanan's (1988) study also suggest that reduced need for skills, coupled with less down time, and less dependence on maintenance staff, due to more reliable machinery, were reasons for introducing CNC. Although it is not clear if these were strategic aims of managers or merely consequences of change.

However, consistent with labour process writers, the intention at Ferranti (Wallace and Whitehall, 1984) was to keep planning and execution clearly separated, to the extent that programmers would have to be called, even to make minor amendments to programs.

Conversely, in the Batstone *et. al.*, (1987) study of CNC, managers favoured some operator control over programming. This was partly because there was no separate programming function in the factory, but also because they wished to exploit the operators skills in programming. The decision to purchase the machinery itself was taken because of an increased workload, and a calculated increase in productivity of 100-200 per cent (which never materialized) based on reduced 'floor to floor time' and the flexibility of the machinery.

These studies demonstrate the presence of strategy in some cases and in other cases there is evidence of *ad hoc* approaches. In each of these, rationales vary. There is little evidence of control motives, it is more common to find that competitiveness, quality, and the technical abilities of the machinery, have attracted managers and informed their decisions. These remain areas that would benefit from further clarification. Beyond these issues it is important to consider who is involved in the decision making processes as this may colour the outcome. Consultation with workers in particular, may yield vital information for the formulation of realistic rationales, and successful strategies for implementation of a plan involving the reorganization of work following new technology introduction.

Work organization

Although there are many variations of the theories, a division can be drawn between two views of effects on jobs after new technology introduction. The first of these supposes a deskilling of work, the second an increase in skills and responsibilities. The pattern cannot be, and is not, nearly so simple as this but it is important to establish general trends since the job outcomes may be informed by managerial intentions. The other issue crucially linked to the division of labour is the allocation of labour which again is an important managerial decision. Allocation policy in turn will affect the work organization and job design.

Five separate processes can be identified in CNC operation beginning with the computer programmer who produces a program from a specification of the component. Following this the machine is set with the correct tools and attachments (in some CNC machining centres the tooling is often controlled by the program that calls up the tools from a carousel attached to the machine). The program is then 'proved out' or put through a 'dry run' to check there are no obvious problems. The next stage is to make any amendments to the program during operation. Equally important is the unloading of the component once the machine has completed its run, as indeed is loading the machine to begin with. The significance of these routines as task components of CNC related work lies in the organization of groups of tasks into jobs. Managers operating within a Taylorist design strategy would retain the tasks within these demarcations, perhaps even seeking to fragment them further as a means of retaining control. The contrary scenario suggested by sociotechnical theorists may be combinations of tasks that devolve greater autonomy to the operators, by virtue of the responsibility embodied in those tasks. For CNC the former would mean, for example, a few manual tasks such as loading and unloading, perhaps combined with simple operating tasks. The latter indicates a fuller job specification perhaps including some programming and trouble shooting responsibilities alongside the other tasks.

The labour process view put forward by Braverman, Shaiken and Noble suggests a transformation and a simultaneous degradation of work. In CNC the expectation is a clear division of tasks in the style of scientific management organization so that the operators of the machines have little skill and therefore lack control over the work. In the five firms studied by Jones (1982) however, the US writers' model, was not sustained. Here there were combinations of setters and operators' functions emerging, and in two firms setters were given staff gradings; while according to the US writers, jobs are becoming more fragmented. Burnes (1989) similarly has highlighted the variation in work organization. This emerges from his findings that some firms had a policy of operator programming, others tolerated unofficial arrangements and in some cases the operators were expressly forbidden from any form of programming or amending.

A significant finding of Jones' (1982) work was that part programming requires a knowledge of tooling performance and metal qualities which is usually gained through shopfloor experience. Hence the ideal candidates for programming functions are setters and operators. Indeed, other studies have demonstrated that managers were aware of this and anxious to exploit the knowledge of the workers (Batstone *et. al.*, 1987). Braverman (1974) also considers the operators of machines as the natural candidates for programming functions. However, where Braverman argues this 'almost never happens' (p199), Jones provides evidence that the involvement of setters and operators, in the amending of programs was a common occurrence, in his investigation.

Findings in the study by Batstone *et. al.*, (1987) agree, but here, skilled operators had insisted and won their claim to handle a full programming function even on complex machining centres. Alternative organization was required of CNC and conventional machines because supervisory staff invariably knew less about CNC than the operators themselves. This left a good deal of discretion both in the way the work was done and in the discipline of completing the work.

Nevertheless, elsewhere in the Jones' studies, there was strong opposition to operator programming and editing in principle, which had nothing to do with ability, but was because the management feared any amendments made would not be recorded. However, this problem could easily have been overcome by ensuring that operators advised programmers of any changes, so that master copies could be amended. This occurred in one firm studied by Wilson and Buchanan (1988). Concern was also expressed about the possible loss of valuable components, and some employers had ordered machines without a shopfloor amending facility. Other firms having such a facility chose to lock it, preventing access by operators. While the organizational politics apparently did not allow operators to get involved in programming, there was clearly an attempt to harness the skills of operators, since some firms were beginning to use apprentice trained people in the planning office.

On the other hand, a strong operators union and a weak programmers union in another firm (also Jones, 1982), coupled with a distaste of the head of programming to get involved with trivial amendments, led many operators to program amendment. Similarly, a different set of circumstances, based on the non availability of programmers on night shifts (and at some times during the day), prompted the overriding of the machines to manual to relieve blockages in production. These informal arrangements are evident also in the work of Burnes (1988) who observed a policy of office programming, but an informal arrangement of operator programming, when production engineers were under pressure. Interestingly, the same study also witnessed a policy in which operators were expected to handle aspects of programming. This was perceived as a problem by supervisors who had little control over this feature of the work since they had been left out of the CNC training arrangements (Batstone *et. al.*, 1987). The supervision of the tasks tended to fall on programmers. However, the supervisors were troubled by this and attempted a reorganization of work around the production engineers so that operator programming passed to production engineering staff in the office.

Another study even demonstrates open negotiation between programmers and operators for the amending work (Wilkinson, 1983). Here both parties wanted the work, although the programmers invariably had the upper hand in making the decision as to whom it was given. The case of the operators was supported by the supervisors who saw programming as a means of compensating for the loss of other skills. The managing director of the company in this study fully recognized the expertise of the CNC operators in programming, which was largely self taught, and wanted to make those experienced in this area into leading hands responsible for proving out and assuring the right settings.

This can be viewed as a Taylorization of the work in so far as it is an attempt to compartmentalize the tasks and isolate certain functions for a small group to perform, instead of using the larger group of operators to perform the functions. Therefore, the negative side of this is the inevitably fewer skills that the other operators would be left with. In short, benefits for one group would mean disbenefits for another (Burnes, 1988). However, from the operators' point of view, the original managerial decision to exclude them from the CNC programming functions, had produced a response of self learning as a means of reintroducing some semblance of skill back into their work. The companies consideration of these people as a potentially new grade of operator would never have been realized without the initiative of the operators. This is all the more significant since the company was prepared to invest little time in the training of the workers for CNC operation.

Wilkinson (1983) found that skilled operators of CNC machinery involved in programming were both a threat to jig and tool draughtsmen and to programmers. Hence, this dimension of job definition demonstrates that it is not a simple decision for managers to decide to have workers doing simplified work or more complex series of tasks. If the latter is selected the functions of other staff may be eroded causing concern for job security and skill status. CNC machines in Wilkinson's main study had been responsible for removing operator tasks in setting and craftsmanship, and these skills had been transferred to the office. However, there was a good deal of sympathy towards operators amongst the foremen and superintendents who supported the notion of operator programming because of their loss of skills. Editing and MDIs were highly contentious issues, and the retention of the keys to CNC computer control cabinets by operators on the shopfloor was against the wishes of programmers. Programming remained the domain of the programmer but some negotiation went on to allow, occasional and unofficial, operator amendment. While in other studies, the union's (TASS) policy to control the hours programmers worked, resulted in operator involvement in amending and overriding programmes by default since programmers were often unavailable (Jones, 1982).

The scenario at Ferranti (Wallace and Whitehall, 1984) began with more rigid control over the programming functions of CNC, ensuring that the only access to these was by programmers or by skilled operators, if closely supervised by programmers. Considerable industrial unrest ensued when the operators insisted on a programming role. The claim for the work was met by resistance from the programmer/planners, and the upshot was a union dispute between the AUEW and TASS. Management solved this problem by giving each side some of the others' skills.

Both Wilson and Buchanan (1988) and Jones (1982) have identified the paradox that Ferranti attempted to solve. There is a requirement for skilled knowledge of feeds, speeds, sequence of cutting and so on. Skilled machinists are aware of these aspects and are therefore ideal candidates for training for programming. The fact that they are not included in, or play only a minor role in programming, may be due to a preoccupation of managers with regulation of the workforce and retention of control (Wilson and Buchanan, 1988). Evidence from other studies does show that some managers are aware of the importance of harnessing existing craft skills and allowing operators a key role in programming (Batstone *et. al.*, 1987). Nevertheless, in Wilson and Buchanan's (1988) studies the turners were themselves prepared to argue their position from this technical standpoint to convince managers of their point of view. In the same study programmers displayed a degree of understanding of operator programming. In one company they would allow programming at their discretion, in another they would leave sections of a program undone, allowing the operators to use their initiative to complete the work.

In other studies there was considerable concern from supervisors feeling that their authority was undermined. This resulted largely from a lack of knowledge of CNC, and an inability, as a result, to control the work (Burnes, 1989; Batstone *et. al.*, 1987). Jones (1982) also found a reassertion of formal channels of communications by line managers. In one company, operators ceased referring problems to them and went direct to the planners. Similarly, there was a conflict between shopfloor supervisors and production planners (both members of TASS) in the study by Batstone *et. al.* (1987), as a result of the latter undermining the authority of the former.

The distinction between NC/CNC and conventional machinery is one that is not readily understood by conventional operators, not trained to work with such machinery, and we can apparently include supervisors in this also. Some studies refer to animosity between these two groups of workers (Wilson and Buchanan, 1988), and others show that managers went to great pains to avoid making a special case of CNC operators (Batstone *et. al.*, 1987) and were anxious to fit new machinery in with the existing company structure (Burnes, 1989). In Wilkinson's (1983) study it was the trade union who objected to the creation of an elite group of CNC leading hands, because of its potentially divisive nature in the firm. Similar trade union concerns about CNC elitism have been voiced elsewhere (Batstone *et. al.*, 1987).

Apart from the demarcation problem with planners, and the undermining of supervisors, the difficulties that emerge between groups of employees are likely to be another potential problem that may frustrate job designs.

In these studies, it is clear that job design outcomes are unpredictable and depend on several factors. It is not possible, on the basis of the evidence here, to argue one way or the other, that there is any particular tendency in metalworking, for a deskilling or reskilling of work. However, any suggestion of a general tendency to deskill work would seem to be refuted by the evidence. Important questions that are raised concern the allocation of work, as much as the job designs themselves. Hence, the decision of how to distribute the tasks between programming staff and operators of CNC machinery, is of key significance to work reorganization in metalworking.

Trade union influence on work organization

The starting point for this section must be to establish what influence trade unions can have. The progression from this issue is then the extent to which they exercise these powers. This in turn is effected by the trade unions local resources. Lack of knowledge amongst shop stewards may restrict the ability to make a case for full inclusion in discussions.

Many critiques of the labour process approach have tended to highlight the lack of consideration of trade union influence in changes. These critiques see the approach as based on the assumption that trade unions are largely ineffectual (Friedman, 1977). This is particularly true of counter arguments to the Braverman theory. Although Shaiken (1975) suggests, US trade unions have largely failed to tackle fundamental job design questions, and tend to be concerned primarily with job security and pay issues. Despite these being American findings, trade union involvement at this superficial level has also been identified in the UK by Wilkinson (1983). Although even here the trade union did play an active role, in their opposition to the creation of an elite group of leading hands, to handle the CNC functions. This response may be interpreted, as the union merely attempting to deal with potential internal conflicts, or the possible loss of members, but could equally be seen as concern over the need to retain the advantages of CNC work for a wider group of workers. However, in Wilkinson's study, the fact that there was trade union intervention occasionally to prevent programmers from performing an editing function, does tend to point to some awareness of job design issues.

Other studies (Jones, 1982; Batstone et. al, 1987) suggest that trade unions can have important influence, given favourable conditions. In one example (Jones, 1982) the AUEW (now the AEU) were well organized and as a consequence able to establish the involvement of skilled operators, who were mainly craftsmen, in areas of CNC programming. The union influence was not the exclusive means by which this was achieved, however. The context also involved a weak section of the TASS trade union (now MSF) which represented the programmers, and a head of the programming section who did not want his staff to waste time on minor tasks such as amending programs. The special circumstances were turned to the advantage of the trade union and the operators, but the scenario was not paralleled in other firms in the study. Elsewhere TASS produced an effective lobby opposing operator amendments of programs. However, at the same time union demarcation rulings prevented direct intervention of programmers.

It was common for operators to override the machines to manual mode to clear blockages (Jones, 1982). At Ferranti a conflict between the TASS and AUEW trade unions over the handling of operator programming prompted the company to revise work organization, and hand editing functions to AUEW operators, at the same time appeasing programmers by giving them some operator skills (Wallace and Whitehall, 1984).

In the study by Batstone *et. al.*, (1987) not only did the trade union insist on the retention of a full programming function by skilled CNC operators, but they also claimed a place in the investment decisions of the company under the auspices of a new technology agreement (NTA). While this proposal for an NTA failed, the company did allow the trade union convener to be party to information, sometimes of a confidential nature, and arranged a new technology review each year. The announcement of CNC introduction brought two main union claims, the first calling for benefits gained to be distributed equally amongst members of the workforce, and secondly, an avoidance of special treatment for CNC workers. The union argued from the point of view that it was important for operators of CNC machines to program if they were to retain their skilled status. Jones (1982) also reports that one shop steward saw CNC as a '....wedge to 'claw back' operator responsibilities.'

However, throughout, the convener had promoted a non confrontational approach, recognizing that the trade union had a responsibility for profitability and CNC could be used as a medium to achieve union aims. Here the sophistication of the AUEW was not equalled in other trade unions on site, which may account for some of their success (Batstone *et. al.*, 1987). For example, Jones (1982), referred to above, has shown that a coinciding weakness in one union and strength in another can determine the success of the stronger union. However, the same study by Jones reports little in the way of any union strategy but, in the company studied by Batstone *et. al.* (1987) the trade union had both a knowledge of the issues and planned for change. Some of the aspects of work organization felt necessary for CNC operation, but differing from conventional operation, such as guaranteed bonuses, were accepted by the union even though they were potentially divisive. This was because it was a union aim to eventually have every worker on guaranteed bonuses. Therefore, this marked an important step in the right direction, and a point of departure for bargaining.

Although some examples of effective trade union intervention in decision making exist, too many trade unions lack the essential knowledge of job design and new technology issues. Agreements in redundancy levels, redeployment, pay levels and the like are common but these are outside any discussion of the work organization itself. Other problems for trade unions are likely where work reorganization gives benefits to one group of workers at the cost of another group, for example between TASS and the

AUEW in Jones's (1982) study. Here unions may find themselves in conflict with each other. Further consideration of the former issue and examination of the latter, would be beneficial.

Case studies in newspaper printing

The newspaper industry with its history of political conflict, ostensibly provides a useful area for studying a process of technological change where labour has had a substantial influence on working practices in the past. Unfortunately there are very few recent case studies that look at work organization and newspaper printing technologies. Those that do fail to focus on work organization issues. For example, Cockburn (1983) looks into the gender issues in the industry, and Martin (1981) deals with the industrial relations issues. Although both have important contributions to make to the more specific issue of work organization.

Equally important is the lack of material on the regional press, Martin, for example, deals exclusively with the national press and yet there is no doubt that, apart from obvious differences of size of the firms, important differences in the levels of control and militancy of the workforce also exist. However, there are useful lessons to be learnt from all of these studies and from international examples such as that of Zimbalist (1979b).

The general changes made to work organization as a result of new technology in newspaper printing are common throughout the industry, to this extent it is possible to generalize. Although, as with CNC, differences arise according to the way sets of new tasks are grouped into jobs, and in the way those jobs are allocated.

The old technology involved the following stages: a reporter would produce a story, or a member of the advertising staff would sell an advertisement and this would be typed. From here the article or 'copy' would pass to a sub editor who would check the work for errors and general presentation quality. Next the corrected typewritten copy would be passed to a compositor who would operate a linotype machine. Linotypes would deliver metal slugs for each letter down a chute and into a galley (or collecting tray) which would eventually constitute a page. From this point a 'reader' would then take an ink copy of the galley and read it for errors. Once correct the stereo department would produce a papier mache mould from the galley into which molten lead would be poured to produce the plate that went on the printing press (Cockburn, 1983; research interviews).

In most cases, the new technology changes involved two stages of innovation. The first stage was the use of a photocomposition system, often followed by a more versatile press to complement the system. In production led photocomposition systems, the change is in the jobs and the means of operation. Strong NGA organization has tended to mean that most of the time the same people remain performing the composing function, although there is commonly a reduction in the number of workers. Nothing really changes for the journalists, but when the typewritten copy arrives in the page make up, area the operators no longer use linotype machines but computer visual display terminals (VDT), and the readers also operate 'on screen.'

The skills for these workers change, the keyboards are conventional QWERTY style, whereas linotypes are far more complex, and the heavy lifting associated with the galleys and metal slugs completely vanishes. The new process is photographic, a printing plate is produced from a photographic negative or bromide of the page (Cockburn, 1983; research interviews).

However, the more significant changes occur with the shift to 'direct entry.' Here the journalist's jobs change, which results in the redundancy and removal of a large number of composing staff and quite possibly the replacement of the remainder. Reporters input their own copy via VDTs, sub editors edit on screen, and from there the bromide and plates are produced. The 'second keystroke' is thereby eliminated and although considerable new equipment is required the changes are more organizational than technological. The most radical technical changes having been made in the switch to production take photocomposition. In advertising, 'tele ad' staff enter advertisement material direct, contributed copy (features articles and reports submitted by freelance journalists) is left to inputers to handle. Inputers may be ex-compositors, but in many examples they are new staff, generally women, with typing skills. They are cheaper and may be employed on a casual or part time basis, so giving greater flexibility (Smith and Morton, 1990). Hence, it is common for existing traditional staff to lose their jobs, in many cases receiving handsome severance payments. There may well be openings for ex-compositors in the preparation of the more complex display advertising, and some papers have retained a 'reading' function.

Managerial intentions

The regional press differs considerably from the metalworking sector. The universality of the new technology changes and their broad sweeping nature, means that a whole newspaper company is affected rather than just one small area or department. The industry wide changes are inevitably informed by a range of factors that are qualitatively different from other sectors. Rationales are thus expected to be more numerous in newspapers than in other areas. One aim is to establish what those rationales are and to what extent they have been informed by planned strategies.

In the past printing has been highlighted as an area, in both Britain and America, in which craft control and autonomy has been preserved (Blauner, 1964). However, latterly, with the introduction of new technology systems, it has also been used to provide an example of managers' use of technology to make traditional printing skills obsolete, with the objective of locating control with the managers (Braverman, 1974; Zimbalist, 1979b). Hence, these US labour process accounts view the printing industry as having '....combined rapid technical change with a reputation for effective worker control over the labor process....' (Zimbalist, 1979b p103). Two examples from the US demonstrate this.

The Washington Post appointed a manager with the brief to develop the capacity to operate without trade union backing. New technology was chosen as the medium to achieve this. Management of the paper had been frustrated by union opposition manifest in sabotage, 'soldiering' and displays of dissatisfaction by, for example, fining anyone who spoke to the production manager, and stopping production if an executive appeared on the shopfloor. At the New York Times, managers even conceded veto power over new technology introduction to the craft unions. However, the eventual outcome was closures resulting in heavy job losses. When the new technology veto was rescinded, the managers were determined to win control over areas that had been so dominated by the craft unions (Zimbalist, 1979b).

In the UK Cockburn (1983) has identified a similar underlying motive of control and refers to the new technology (production take photocomposition) as 'the weapon with which to smash the compositors....' (p61). Smith and Morton (1990) have provided more recent evidence on union exclusion policies. Although often there are considerable financial benefits in this course of action as well. The examples provided, in Martin's (1981) case studies of national newspapers, identify a poor financial record as a primary reason for the consideration of new technology. Times Newspapers Limited (TNL) (Martin, 1981 p254ff; Cockburn, 1983 p79ff), for example, were under considerable financial pressure from falling advertising and circulation revenue, and decided on the use of photocomposition systems to alleviate the crisis. This was to be achieved by the elimination of dual keystroking (typing the text twice) which would be reformed piecemeal, starting with the Times Higher Educational Supplement. The primary aim was to reduce production costs. Mirror Group Newspapers (MGN) also had significant economic concerns, and were in the process themselves of looking into reducing production costs by reducing staff. The use of photocomposition and electronic page make up was to play an important role in achieving this. Similarly, the Financial Times (FT) (Martin, 1981) were anxious to use new technology as a means of reducing labour costs, again by facilitating lower manning levels. This system was ambitious, new computers were to be set up in production, editorial and the commercial areas of the business and the plan was to be implemented in eighteen months.

At the regional level Cockburn (1983 pp71ff) and Smith and Quinlan (1982 pp14-16) studied the Croydon Advertiser. Here reasons for moving to a photocomposition system were related to the need to increase capacity at the newspaper, but also because parts were becoming increasingly difficult to obtain for the old linotype machinery. The newspaper also saw the introduction of new technology systems as an opportunity to close the gap between office staff and production staff, who were traditionally separated from each other. King and Hutchings, another regional newspaper, introduced new technology after their premises were gutted by fire. Later they also introduced more new equipment as a second phase in the programme (Cockburn, 1983 pp68ff).

One further regional study (Smith, 1988 pp214-221) has identified the retention of managerial control, as an important element of new technology usage. However, this newspaper (the BPM) had been involved in other reorganizations previously with a move in production sites. The changes were intended to improve the response of advertisers, and they suggest that there was concern about the newspaper's market position. '....from 1979 onwards the increasingly competitive and declining market for advertising and circulation compelled the BPM Ltd. to remodel its product range, rationalize production, introduce new machinery, negotiate new collective agreements and to redesign the division of labour as a whole' (op. cit. p219). These particular courses of action were determined by an equally varied range of issues:

Management's decisions were determined by their perceptions of the savings in labour cost, the potential impact on the contracting market share of the company, the availability of management skills, and the degree of opposition expected from the labour force (Smith, 1988 p219).

Managers were influenced by trade union opposition because, on rejection of direct entry technology by the NGA, managers sought alternative cost saving measures in reducing working hours and jobs in the company as a whole. However, the fact that considerable opposition was experienced from the production workers, turned managers attention to the editorial workers. Here cost reductions were achieved when different editorial teams were amalgamated taking advantage of weak union organization in this area.[4]

In newspaper studies the weight of the evidence suggests that motives of control over workers are common. This is stimulated by the reassertion of managerial prerogative, based on an historical distaste of trade union control. However this is inextricably linked with a need for new technology, in line with competitive pressures, and due to the increasing non availability of traditional forms of new technology. However, this does not exclude other motives of workforce integration although these are rare. Any definite strategy is difficult to establish but because approaches are so similar it suggests that rather than strategic, the procedure is an established pattern. Clarification is required to confirm the universality of rationales, to consider the influence of workers to these aims, and to establish the planning procedures.

Work organization

An analysis of the work organization outcomes in newspaper printing, after the introduction of new technology, are important to establish the resulting job designs and to consider the allocation of work. Given the existence of evidence above, of rationales for the control of workers. Here political considerations, in some cases, seem to have effected both the design of work and decisions about who was to take on the work.

The studies of work organization in the printing industry have tended to refer to the era of production take photocomposition systems introduction. However, the direct entry changes made at the Times gained considerable notoriety. TNL was bought by Rupert Murdoch's News International Group and a new site at Wapping designated for a full direct entry system:

> By moving to a new 'green field' site Rupert Murdoch had succeeded in the most dramatic fashion in introducing new technology on his, rather than the printing unions' terms, something other Fleet Street proprietors had failed to do for over a decade (McLoughlin and Clark, 1988 p2).

Murdock's success has completely changed the course of industrial relations in printing, and moved newspapers out of Fleet Street. Nevertheless, existing analyses of attempts to introduce production take photocomposition technology into Fleet Street 'nationals' are far from being purely historical accounts. They demonstrate possible difficulties that may be encountered in other strongly unionized areas.

Martin's (1981) account of the earlier changes attempted by the Times (under the management of Times Newspapers Limited (TNL)) details a traumatic industrial relations crisis in which ultimately, the management were able to achieve little of their original plan. Agreement was reached on introducing photocomposition but the NGA retained the inputting function, the only inputting that was allowed by other staff was very limited. However, management did achieve greater internal flexibility of the staff within the production department. Mirror Group Newspapers also had ambitious plans that were not realized. Some cooperation was achieved to introduce photocomposition, but like the Times the NGA hung on to its control of keyboarding.

The report on the West Midlands regional paper (Smith, 1988) centres on union activity, little is reported on the changes in the work. In fact a substantial number of traditional skilled staff, mainly compositors and readers, were lost, but around thirty per cent of these went forward to train as sub editors.

The importance of this redeployment is in the recognition of the skills embodied in the workers, and at the same time acknowledging that it is a realistic proposition to retrain the existing staff rather than introduce new staff (Smith, 1988).

Unfortunately Zimbalist (1979b) is too preoccupied with control issues, and the way the context in which job changes were organized to force out the trade unions, to deal with the way the work was actually reorganized. Although it is implicit that new technology's introduction would involve deskilling of the work along the expected lines. Here this is to specifically facilitate removal of the skilled union employees, therefore establishing a more detailed division of labour. Cockburn (1983) does deal with these issues in respect of production take photocomposition. She argues that deskilling of the work results from the ability of the computer to perform tasks such as hyphenation and justification, tasks that would previously have required considerable skill in manual operation. Inputting becomes a less precise art, also general quality control procedures of checking work were no longer required. These aspects of work not only withdrew skilled aspects from the work, but perhaps more importantly, the work was no longer so highly skilled as to require such extensive training. Cockburn (1983) asserts that 'Composing has lost its mystique....' (p107).[5]

As Cockburn (1983 p95) has argued 'often a new labour process implies a new labour force.' However, in the majority of cases in newspapers, and certainly all the cases presented here, the new labour process under production take photocomposition did not bring a new workforce despite the significance of the changes. The new workforce arrived, in many cases, with direct entry technology, fashioned together by new workers, retrained compositors and journalists inputting copy direct. The success of the compositors' resistance to change, was primarily because of the strength of their trade unions, but this strength was firmly underpinned by their exclusively skilled status. For Cockburn (1983 p113) skill is conceived in three ways. It is accumulated in the individual, it is demanded by the job and it is a political construct used by workers and trade unions to defend areas of craft work. The majority of the skills, on which craft status was based, became redundant with the change to photocomposition although, the loss of skill in the job was buttressed by the political defence of the craft skilled status.

Hence the problems encountered by managers stem from the strength of the trade unions, particularly the NGA, which not only stood in the way of new technology for many years, but also guarded its skilled status in the industry very jealously. Again Cockburn shows how this itself eventually created a management imperative for technological change that would destroy this power base founded on skill and craft: '....craft skill represents a constraint on managerial initiative which may became intolerable in certain economic conditions' (Cockburn, 1983 p113).

Economic conditions undoubtedly did play an important role in many cases of new technology introduction, but even with these external pressures, the entrenched position, and power of the print unions, has the potential to inflict injury, not only on the business and managers, but inevitably on its own members.[6] Hence, recent history has shown that eventually something has to give, and the trade unions that refused to compromise are broken one way or another.

In the aftermath, the problems for managers are how to deal with the workforce that, after new technology introduction, is largely superfluous. They can, and do, redeploy staff as in the Smith (1988) study where composing staff were retrained as sub editors. However, it is noted that *en masse* removal of NGA craftsmen may well be favoured for reasons of cost reduction, avoidance of 'equal pay for work of equal value' claims, and to avoid future union organization. Smith, however looks at both photocomposition and direct entry changes, while the majority of studies only deal with production take photocomposition. In most cases dealing only with the latter, the production of the newspaper was left in the hands of the skilled craftsmen who are reported to have found the new work undemanding and not satisfying.

Locally the Croydon Advertiser's (Cockburn, 1983; Smith and Quinlan, 1982) NGA employees lost their traditional craft skills in the change to photocomposition. The new skills they acquired were not considered as demanding or requiring the same ability as linotype machinery. Increases in pay softened this blow and the change was accepted as a reluctant compromise. Similarly, at the Daily Echo (Preece, 1987 pp3-8) workers were unhappy about the loss of skill, but the managers responded practically with job rotation to provide more variety. This was in addition to pay allowances and reduced hours intended to motivate the workforce.[7]

An analysis of work reorganization in regional newspapers reveals what seems to be an inevitable deskilling of traditional workers, with the introduction of new technology. In the majority of cases this also means a significant loss of jobs in the pre-press area,[8] and the employment of new workers. However, the retraining of compositors for redeployment in other parts of the company demonstrates another possible allocation scenario. Variations in work organization are possible but the limited number of existing cases makes analysis of new studies useful to either confirm existing general trends of deskilling, or to challenge these. At the same time an analysis will reveal allocation policy, particularly the use of redeployment, as an alternative to job loss for traditional workers.

Trade union and worker influence

Given the tradition of trade union control in newspapers, such as the pre-entry closed shop in the UK, one of the most important issues is the shift in the power balance between trade unions and managers after the introduction of new technology. The first issue, of key significance, concerns the conduct of this transition. Since union power was based on craft skill and autonomy, the introduction of new technology reduces the need for traditional skills, and inevitably, erodes the union power base. Against this background, another important issue concerns inter union (NUJ and NGA) conflicts and cooperation over impending changes.

The operating environment for most newspaper companies has generally been against strong union opposition. Here, there is significant evidence that managerial control motives exist, where managers wish to regain control over fundamental areas, seen by this group as managerial prerogative. This means, for example, the breaking of the pre-entry closed shop, and the overall suppression of the NGA's power to dictate terms to managers, by means of the various forms of industrial action. Hence, the control motive is present in different measures in the industry, understandably it is one the trade unions often highlight.

Zimbalist's (1979b) Washington Post study provides a good example of where managers had set out to undermine union control, by training non union workers to operate the machinery. The capability of this group of staff was proved when they continued to produce the newspaper while the printers were on strike. The use of new technology systems aided them. When the printers returned to work they sabotaged the press to prevent production of the newspaper. The 'Post' responded by contracting out the printing of the paper. Ultimately the paper handlers reached agreement against the printers. This was an inter union conflict sparked by a previous dispute involving their non admission into the print room by the printers. The paper handlers were then offered print room jobs, thus enabling the management to remove the skilled printers altogether.

The unions at the New York Times felt a compromise over new technology was essential since the newspaper was in financial trouble and as soon as the new technology was implemented the consequence was a loss of jobs of many of the skilled staff.

Times Newspapers management (Martin, 1981; Cockburn, 1983) adopted a 'macho style', involving management insistence for the rapid conclusion of agreements on the continuity of production and new technology. However, the working parties looking into new technology introduction specifically excluded trade unions, and these were the forums that would have discussed job redesigns, allocation and redeployment. This led to worsening relations. The NGA refused to negotiate on a number of issues. These included new staffing arrangements and wage structure, cessation of restrictions on production and unofficial action, and the introduction of new technology.

Since the union was not forthcoming, management closed down the newspaper for one year. When the newspaper reopened, it was in recognition of the fact that negotiation could not take place during a period of closure. Agreements were reached that there would be no compulsory redundancy, and that adequate training would be ensured. Agreement was also reached to introduce new technology, but this was to be strictly within existing demarcations. Besides this, few concessions were made by the NGA.[9]

Unlike TNL, MGN recognized that the industrial relations issues were really located at chapel level,[10] and began by adopting close consultation and attempting to encourage inter chapel cooperation. Pay rises were agreed in return for cooperation and photocomposition was eventually introduced, but as at TNL, keyboarding remained firmly in the control of the NGA. The Financial Times had similar experiences to both TNL and MGN, but particularly experienced rifts between unions. For example, the NATSOPA was much happier about altering demarcations than the NGA. Although it was recognized that a national and house agreement was of key importance, lack of union cooperation, and a poor consultation procedure, meant that important knowledge of the production process was unavailable.

The Croydon Advertiser provides something of a model case, in terms of the provision made for consultation by the management. This included the opportunity to comment on the type of system to be used, and the training that would be utilized. The workforce actually rejected both of these opportunities, fearing that their bargaining position may have been weakened in these areas. Eventually it was agreed, that there would be no compulsory redundancies, and there would be considerable additional payments for operating new technology. Here retraining of staff and the implementation of new systems was handled carefully (Cockburn, 1983; Smith and Quinlan, 1982).

Consultation on the technical issues, as well as the new working arrangements, included the union representatives and supervisory staff at the Daily Echo (Preece,1987). Although here, the discussions tended to centre on issues of compulsory redundancy and the financial enhancements that the workforce would enjoy, in recognition of their acceptance of new technology. In the King and Hutchings organization, the union was not actually consulted about new technology introduction, but a fire at their premises led to the automatic replacement of conventional equipment with a photocomposition system. This was implemented outside of any union agreement, but when the managing director changed, agreement was reached and more advanced equipment was introduced later, with the eventual result of job loss and wage reductions due to a contraction in business (Cockburn, 1983).

Smith's (1988) study in the West Midlands reveals similar patterns in the maintenance of the division of labour, with the change to photocomposition and direct entry. The NGA were given concessions to secure their support for the changes, but initially the division of labour was maintained. However, the proposals for direct entry were rejected by the NGA, so the newspaper sought other areas for cost reduction, and proposed the integration of the editorial teams working on separate newspapers. Eventually the NUJ and NGA agreed to work together (under the 'Accord'). Management ordered the direct entry technology and the NGA, still in opposition, decided against strike action but set about achieving an agreement on the issue of redundancy and redeployment. Forty compositors and readers lost their jobs but twelve of these went forward for retraining in sub editing.

In the printing case studies, union strength based on traditional skills gives way under the pressure of new technology systems, which enabled the same functions to be performed by lesser skilled workers. The lack of adaptability of the NGA and managerial distaste of the union, in many cases, led to minimum inclusion of the NGA in discussions. Consultation and negotiation was generally restricted to redundancy numbers, levels of pay and redeployment arrangements. Very little evidence exists of job design or work organization discussions. New studies should assist in a better understanding of union involvement.

In one case, potential conflicts likely between the NUJ and NGA were avoided by agreement for cooperation. This was the response to the newspapers' transition to direct entry, little research exists considering this particular aspect of new technology introduction. Hence, some clarification in this area is of particular importance.

In the US and in the UK (IRRR, 1985) there is evidence of trade union conflict based on the preservation of existing demarcations. Analysis in new studies should serve to clarify this issue.

New technology in the office

New office technology is more varied than the kinds of technologies discussed in the previous sections. At the more simple level it includes word processing and database systems but, increasingly it is being used by all administrative and clerical departments. For example, in buying and inventory control, which may have direct links with the manufacturing departments and provides a reliable source of information and communication. Many of these departments have had the use of a mainframe computer for sometime. This is slower, and less flexible than newer systems that are independent of a central computer, but integrated within a wider network of computers.

Even beyond this, there are more and more users of electronic mailing systems, electronic funds transfer at the point of sale (EFTPoS) and the use of bar coding, all facilitating the more rapid transfer of information.

Like engineering and printing, this can have varying consequences for workers, depending on how the component tasks are formed into complete jobs. Many applications will automate the more mundane tasks but some of these always remain. For example, the inputting of data has to be done at some point and this task may be separated off as a job by itself, or integrated into a more complex set of tasks altogether. The former is likely to mean some highly detailed jobs and some more skilled jobs, the latter would probably ensure all jobs had some element of skill.

Another feature of new technology in the office environment, is the tendency for the union organization to be weak or non existent. There are, of course, some firms that have no union, either in the clerical or manual areas, but the tradition of the unionization of manual work is not paralleled in the office (Price, 1983). Where there are clerical and administrative trade unions, they seem prone to lower levels of militancy which is probably the result of the more paternalistic style of management that prevails. The 'staff' classification often brings with it much greater rewards than are enjoyed by their counterparts on the shopfloor (Blackburn and Prandy, 1965).

Managerial intentions

As in the metalworking and printing sectors, the aim of this section is to examine existing accounts of new office technology introduction. This is to establish the rationales for introducing particular forms of new technology, and to consider how carefully the procedure of introduction has been planned. Unlike printing, strong traditions of unionization in most clerical occupations is uncommon (Price, 1983). Therefore, rationales to reduce union control are unlikely to be central. The nature of office systems as sources of rapid information transfer means that competitive edge, information quality and managerial control are likely motives.

Once again, the notion of deskilling and job fragmentation, as a result of new technology usage, is evident in a study by Crompton and Reid (1982), where elements of control have also been removed from clerical workers functions. Other writers have also discovered this degradation of jobs (Glen and Feldberg, 1979), seeing clerical work largely controlled by the requirements of the machine. This style of design prevails despite the workers having the ability to perform more skilled functions.

Investigation of three insurance companies (Storey, 1987) at varying levels of new technology introduction, discovered widely varying accounts of the motives for the use of new systems. The author, in this case, felt the key reason to be cost reduction via direct entry of data, which would reduce clerical staff. Equally, it would mean the facilitation of the merging of branch offices, which would also reduce the more highly paid managerial grades.

In one of a series of studies by Wainwright and Francis (1984), a college of higher education (pp73ff) had plans to introduce word processing for use by secretaries and academic staff, and with capacity for further development later on for data processing. The objective was to increase the efficiency of the college secretarial services.

In a nationalized corporation, the same authors (1984 pp99ff) found that the reason for the wish to introduce new technology, was an efficiency and productivity drive. Here, the systems to be introduced were word processors and electronic filing and mailing systems. The objectives of the system were listed as profitability, viability, competitiveness and to improve organization/managerial effectiveness via improved facilities. Another study, in a financial services company (op. cit. pp130ff), had a policy of automating the more mundane of its office functions. This was coupled with the more general motive of keeping costs down. Computer and telecommunication systems had already improved the companies attainment levels, but the specific objectives of the new word processors were to reduce the cost of each document, improve response times, and maintain quality. Filing and mailing systems were to speed up the retrieval and transmission of information.

The objective of new technology introduction was to improve client service in another study (op. cit. pp162ff). This was to be achieved, in an international insurance broker, by developing more efficient methods of processing and supplying information. This would, in return, give higher productivity in typing and other areas of work, but also facilitate the faster collection of payments, and the presentation of an efficient, up to date image. Here, the company applied integrated document distribution systems, to enable national communication between offices in the UK.

Pressure on managers to reduce costs was an important factor in another insurance company (Batstone *et. al.*, 1987). The increasingly competitive nature of the insurance business, and the use of new technology by other companies sparked a drive for greater cost effectiveness. Although not wholly responsible, substantial job loss took place following new technology introduction, accompanied by reorganization of the company. The strategic approach of managers involved the use of pilot studies, which revealed a series of problems in the prospective changes, making the process of change very slow and incremental.

There are a number of similarities with the engineering companies. Rationales are based on a range of factors, including competition and efficiency criteria and there are some cases where control is clearly a motive. However, often in all of these companies, the necessity for more effective communication is driven by the fact that the competitors use, and benefit from, similar new technologies. This provides the impetus for other firms to adopt similar systems. However, there is little doubt that the advanced nature of some of the systems, facilitates closer managerial control of the work. There is also evidence of some careful planning, and a use of pilot studies prior to full introduction of some systems.

The possibility of there being central motives of control through new technology remain, and require further investigation. The generality of strategy is also unclear. Clarification of these issues through new case study analysis will help to answer these uncertainties.

Work organization

This analysis examines the presence of any general trends occurring in work organization, following new technological change. Secondly, it considers the allocation decisions, of whether to use new workers in new jobs, or whether to redeploy or retrain existing employees.

One area of clerical work in a local authority (Crompton and Reid, 1982), involved routine form filling and data preparation, prior to entry into the computer. This only left a small area of discretion in queries received from payees. It proved to be an environment in which the computer dictated the pace and the volume of the work. Employees were dissatisfied at the lack of personal contact involved, but also complained that they were only taking part in a small portion of the process. In fact, they would have liked to follow the operation from the beginning through to the end. One section leader likened the work to an assembly line, and suggested that staff motivation was very difficult to achieve. Also evident was the use of an unskilled female workforce, who by virtue of their high rate of turnover, provided flexibility with down turns in the organization's work load.

The American study findings (Glenn and Feldberg, 1979), show that, after the introduction of new technology, the possibility exists to create jobs with a higher skill requirement. Here posts were generally filled by new skilled staff, not existing staff subjected to retraining. Existing clerks were left with the more routine clerical jobs and a hierarchical system of control, suggestive of work organization in a Taylorist style.

In the insurance case studies (Storey, 1987; Batstone *et. al.*, 1987), computers were introduced for data storage and to enable the direct entry of data, as opposed to a system of submitting forms for punching onto a mainframe computer at head office. Storey found that little advantage was gained in speed, since clerks were not trained typists, and their typing was only a little faster than form filling. However, what it did mean was that the work could be carried out at branch level without referrals to head office. Other jobs were deskilled because the calculations that would have been done by hand became computerized, and communication between managers and underwriters ceased as the information was increasingly carried in the system. Alongside these changes, a reduction in the need for training was expected, since lesser qualified staff were likely to be required in future. Like the cases above, this example suggests a deskilling and Taylorization of the work.

Reorganization to reduce the number of policies on offer, in the Batstone *et. al.* (1987) study, improved the prospects for the successful implementation of 'on line' processing. As in the Storey (1987) investigation, this meant less paperwork in general, less use of manual files, and less arithmetical calculation. Some aspects of the reorganization potentially add up to deskilling, but instead of handling specific policies, as before, underwriters now had to handle the full range of policies. This involves an increase in skill, and the job changes were accompanied by grading increases. Moreover, the new system failed to completely sweep away the use of manual files, and there was still a need to liaise with clients and maintain manual files.

In the college of higher education study (Wainwright and Francis, 1984), word processors meant that the typists found it easier to cope with their work load. This involved the routine tasks of typing standard letters, examination results, the daily menu, and other copy typing tasks. Those involved, reported greater work satisfaction and less stress, because the routine work could be moved faster, but also because the typists had been given extra responsibilities in the form of managing their own database. This consisted of deciding which files to delete, which to keep 'on line', and which to archive.

For the secretaries, there was considerable job expansion; the responsibility being shifted to them for managing resources for both staff and students. They also took responsibility for various other tasks that were not computer related, but these were areas where the new technologies had clearly given the opportunity for greater freedom. The system itself allowed better control of resources and in future would be employed for stock control and accounts. For the Wainwright and Francis (1984) the case study '....supports the view that deskilling and routinization of work are *not* necessary accompaniments to new technology' (p98: emphasis in original). More importantly, for this book it provides a good example of choice in new technology. Demonstrating that it is a medium that can be used to enhance work and not only degrade it.

However, opposing examples are evident in the nationalized corporation studied by Wainwright and Francis (1984 pp99ff). The computing director felt that word processor operators would be deskilled since the computer software left few areas for discretion. Spelling was handled by the package, mistakes could be easily corrected and, with time, managers would correct their own documents. For this reason, the director had begun to allocate work to personnel trained as key punch operators, on the basis that they were used to routine work, and the requirement for key punch people was receding. Although the computer director's three typists (one an ex-key punch operator), did have a considerable amount of typing included in their jobs, there were also a variety of other tasks such as using electronic filing and mailing facilities, as well as answering the telephone and dealing with direct enquiries.

However, in a subsidiary organization, two contracts and estimating typists had far more intensive jobs. They were required to work with VDUs for long periods without a break, and the local computer manager had plans to create an environment of even more routinized and detailed work. What he proposed was the use of files of standard clauses which would simply be inserted into the appropriate quotation, tender or contract. The contrast between the two organizations is interesting. The same technologies are employed, but the tasks are composed quite differently and this provides another good example of the opportunities for choice in work design.

The financial services company's (Wainwright and Francis, 1984 pp130ff) main new technology introductions were word processors effecting typists and secretaries in the firm. However, the use of the technology had different effects on each of these jobs. The secretaries enjoyed a degree of discretion, since they were working with varied documents, and were allowed to choose whether a typewriter or a wordprocessor was a more appropriate tool.

Originally the typists' work had involved typing documents coupled with a liaison role. This involved contacting the originator of the work if there were any problems with a document. With new technology, the work changed to involve far more detailed tasks and less discretion. Instead, the originators of documents would identify standardized paragraphs for inclusion in a letter or document, which would then be returned to the originator for amending. Direct contact of the kind employed before was discouraged.

Other features that suggest a Taylorized environment were the closer supervision these staff were subjected to, and the allocation of the work. Even within the production of these 'pattern letters', different grades existed according to the level of sophistication involved. Control clerks logged the speed at which individual typists produced work, which was used for salary reviews, as well as a means of gauging overall throughput. Most of the features described in the study, represent classical Taylorist organization, in which discretion is almost non existent and tasks are fragmented in the extreme. The machinery does not have the capacity to pace the work, but this is handled by close managerial control and the use of a (negative) incentive scheme to encourage faster work.

The final case dealt with by Wainwright and Francis (1984 op. cit. pp162ff) is that of an international insurance broker. The new technology introductions here were a network of processors in regional offices reporting to a mainframe computer at head office allowing both the transmission and the integration of information. In the head office group of operators, team working in the preparation of documentation, and the operation of the system was emphasized.

However, this contrasted with a regional office, at which there were two main groups of operators. The first preparing the data for entry, and the second feeding it into the system, thus fragmenting the work to a greater extent than at head office. This was supported by stronger, direct supervision, as compared to a good deal of autonomy at the head office. In another office using the system, secretaries found themselves with considerable discretion and reported directly to managerial staff. The managers were regularly absent from the office, by virtue of their jobs, and this often left the secretaries to deal with client queries.

There is tremendous variance of work organization in these studies. In some, motives of control dictate a detailed style of job definition. Others reflect much more varied work, in which new technology introduces new skills and responsibilities. It is interesting that the same new technologies can generate quite different forms of work organization. Thus, providing good examples of the social and political dimension of new technology. Work in most cases is allocated to existing workers, although some suggestions remain that new workers are used in cases where work is upgraded.

Clarification is required of the tendency to any general outcome of work design, and secondly to consider how the work is allocated, in cases where new jobs and tasks are created.

Trade union influence on work organization

There is a clear qualitative and quantitative difference between blue collar unionism and white collar unionism (Hyman and Price, 1983). Given that this is the case, and also that white collar workers generally enjoy better working conditions than their blue collar counterparts (Blackburn and Prandy, 1983) it might be expected that unions are less militant, or concerned less, with intervening in managerial decision making. At the same time managerial initiatives to discuss issues with workers, with whom contact is easier, and within the same environment, would be likely to be more common. However, union conflicts might be expected where new technology enables a controlling role, which is to the detriment of supervisory workers, but to the benefit of clerical workers.

A common criticism of labour process writers, is their tendency to see management as an omnipotent force through which Tayloristic job designs are imposed on the workforce. Braverman (1974) is particularly prone to this, but Crompton and Reid (1982) and Glen and Feldberg (1979) also seem to assume a correspondingly dominating management. However, this tends to overlook the influence that workers and unions may have on the design of work. Some writers consider concessions to workers in the nature of the work they do as a sop (Braverman, 1974; Ramsay, 1985; Zimbalist, 1975; Dickson, 1981). But it is, nevertheless, important to give consideration to the attempts of trade unions to influence management, particularly in respect of work organization.

In the insurance companies, studied by Storey (1987), two companies were represented by the Association of Scientific, Technical and Managerial Staffs (ASTMS) and the other company by the Banking Insurance and Finance Union (BIFU). These trade unions had negotiating rights, but only in one case did the head office include a union representative in consultative forums. Forums were mainly where user managers aired their views. The staff themselves were not consulted directly. There was an effort to give information to the staff, perhaps more an attempt to get the staff used to the idea of the changes, since, as the new technology was introduced, the information sessions receded.

Significantly more opportunity for involvement was enjoyed by the trade unions at the college study. Here a team was set up with the trade union and central administration department, the computer specialists were not involved in these discussions. Even at the implementation stage the staff were encouraged to develop their own best ways of working the system. The result was a satisfied workforce with greater involvement, interest and skill (Wainwright and Francis, 1984).

Predictably, the same consideration was not given to staff views at the nationalized corporation (Wainwright and Francis, 1984). Here the director of computing made the decisions, and his concern was acceptance from managers and secretaries. Trade unions were not considered to be of any importance in these proceedings. The approach was an attempt to convince secretaries of the benefits of word processing and electronic communications. A team of three secretaries was set up to provide the training, technical decisions were made by technical staff, but the choice of who should be trained, was made by secretaries. The system users were encouraged to develop their own best working practices, according to the demands of the areas in which they worked. While secretaries had a good deal of control over work design, typists clearly did not. However, the typists seemed happier on the whole, and felt that they had more responsibility, but they were dissatisfied that a promised pay increase had never materialized.

Not unlike the nationalized corporation, the financial services company revealed a management attitude, that saw consultation about the jobs and the technology, as time wasting and confusing. As a result little effort was expended in discussions with the users, something which the users themselves found disturbing. No additional payments were offered or made, clearly the fact that there was no union in this company, was an important issue in the managements' approach. There had been an overall increase in the number of tasks, but the workers were concerned about increasing stress, due to close supervisory control, machine pacing, and the timing of jobs. Typists found themselves particularly controlled by this regime, while secretaries managed to retain a greater degree of discretion in the way they organized their jobs. The company had seen an alternative design involving a more participative approach, but had rejected this as inappropriate.

Once again a considerable contrast is evident in the example of the international insurance broker. All future plans were discussed in an office system development group, which was designed to provide '...a focal point for user contact...' (Wainwright and Francis, 1984 p169). The group was conceived as a result of a recognition that there were previous misunderstandings of the technical requirements. The approach was very structured, and took into account users views, from obtaining initial approval for the system, through to deciding on the specification, implementation, and training. This consultation took place in conjunction with a pilot study. People did not leave the training programmes until they were confident enough to do so. Work design itself was not decided in such a planned manner but, to a degree, operators were able to reorganize their work, post implementation.

The study by Batstone *et. al.* (1987) documents trade union (BIFU) attempts to be included in the pilot studies, and monitoring schemes, prepared for new technology introduction. The management admitted this, although in practice, actual involvement was limited. However, the union used members involved in the new technology 'pilots', to monitor the progress of the scheme and report back to them. This again met with limited success since many members failed to report back to the trade union representatives.

An attempt to strike a new technology agreement was rejected, but the union did successfully influence agreements on redeployment and redundancy issues. However, even here this required the pressure of threats of strike action. Guidelines for VDU operation were set, but only adhered to by half of the managers, and issues such as work allocation were management decisions only.

Hence, union influence was limited, and the discussions between managers and trade unions were described as 'consultative rather than negotiative' (Batstone *et. al.*, 1987 p184). A well organized union at company level seemed to be frustrated by a number of problems. These included an intransigent management, local level union ineffectiveness, incomplete membership, and an attitude of workers, towards the union, as a means of protection rather than a possible source of influencing change.

Union involvement seems to be influenced by three issues in the clerical studies. The trade union is not interested or aware of how to deal with the changes, secondly, trade union interest meets with a managerial wish to exclude trade unions from discussions. Finally, in some cases managers openly encourage union intervention. In these studies there is a tendency to bypass the trade union and go direct to the employees and users of the technology. Examination of new case studies may allow more of a basis for generalization. Little evidence of inter union conflicts, due to demarcation breaching, has been found here and this is an issue that requires further examination.

Conclusions

This chapter has considered a number of existing case studies to examine how far the questions raised in chapter two are answered. This has been done under the three headings of managerial intentions, work organization and trade union influence on work organization. The final part of this section considers the questions that remain unanswered and prepares a series of hypotheses for testing against new case study data.

Managerial intentions

In the studies of NC/CNC, labour process writers argue deskilling strategies exist as a means of gaining control. Other studies have argued that plans are not sufficiently well defined to achieve this aim, and much decision making is dealt with on an incremental, *ad hoc* basis. Detailed planning for change is rare. Rationales for introducing new technology also subordinate deskilling and control motives to wider aims. These actually range, from fitting in with the current organizational environment, to remaining competitive, or acquiring a competitive edge. Some studies, contrary to the deskilling thesis, wanted to draw upon operators' programming ability (Batstone *et. al.*, 1987).

Contrary to the variability of aims and outcomes in the metalworking studies above, considerable evidence exists of attempts in the newspaper industry to use technology for purposes of deskilling and winning control over the trade unions. This, sometimes hidden, approach occurred by virtue of the NGA's 'strangle hold' on the industry. Despite this, many of the newspapers attempted to use new technology as a medium for remaining competitive and were careful in the early stages not to present direct challenges to the NGA. Nevertheless, many of the subsequent attempts to introduce new technology were aborted or modified due to union pressure. Following these experiences, managers in both the national and regional press, seem to be attempting both a solution to their labour problems, and a simultaneous improvement in the competitive position of the newspaper.

Some strategic approaches in the clerical studies are mixed with far fewer planned approaches. Here the nature of changes proved to be rather different. Generally rationales, which included reducing costs, required some planning to test their feasibility. Most rationales had no direct deskilling intentions but included aspects such as the reduction of response times, and improved quality of service which was achieved by improving communications.

Work organization

Upgrading of workers in the form of conferring staff gradings and creating setter/operators from operator grades, appeared to be common in the metalworking studies. However, there are a number of examples of deskilling also, but it is not possible to claim that either trend is universal or even general. The way that work is allocated in metalworking, proves to be an important theme in issues of work reorganization. If utilization of skills was the aim, then operators would appear to be the ideal candidates for programming with their 'on the job' knowledge of feeds and speeds. However, their potential trainers, existing programmers and production engineers, often refuse to train because of the perceived threat to their own jobs. Other problems of work design, lie in the often negative view of CNC operation, as less skilled than conventional operation. Similarly, supervisory grades feel threatened because they are untrained in CNC and are unable to give advice on this kind of work as a result.

Significant job changes are common in both the regional and national press. These formed two different kinds. The first was a considerable deskilling of the work, and involved changing the type of machinery, from traditional mechanical linotype machines to a computer photocomposition system. The nature of the work changed radically, but the workers doing the work did not. The second change to direct entry, meant completely new unskilled workers were introduced to handle the, now deskilled, inputting function. Major changes have therefore taken place in both new technology and in working practices.

Like the metalworking studies, work allocation problems occur since there are a stock of old, redundant, workers and a range of new jobs, that rely on an alternative set of skills. In one study this was partly resolved by redeployment and retraining in sub editing posts.

In the office studies, there was a tendency towards varied work reorganization, that reflected a combination of increasing work complexity, on the one hand, and simplification of work roles, on the other. Typical changes involved the use of computers to handle records, but also to deal with some of the technical aspects of the work such as calculations. The handling of paper was greatly reduced, and communications across short distances and large geographical areas are much improved. Some discretionary parts of the work tended to vanish, particularly personal contact, since better information retrieval reduced the need for verbal communication. However, alongside these changes, work loads are often increased to include a wider variety of tasks. Nevertheless, there were a considerable number of examples of highly divided work.

Trade union involvement in work organization

Contrasting evidence, demonstrating both the potential effectiveness of trade union organization, and conversely, ineffectual activity, existed in the metalworking companies. In some cases, poorly organized or weak trade unions, failed to satisfy managers that aspects of programming should be included in operators' work roles. Other cases demonstrate that unions are able to influence managerial policy on these issues.

The trade unions in the clerical organizations, like those in the metalworking companies, were able to influence management decisions in a number of cases. However, in other examples involvement was limited even where unions were well organized. In the latter cases, it seems clear that without management agreement in the principle of involving unions, the trade union's case for inclusion becomes stifled.

Invariably, concessions were possible from newspaper managers, where the NGA had accepted that new technology changes had to occur. Where trade unions refused to accept this there are examples of frustrated managers sacking the production workers, and employing new workers to handle inputting functions, as well as adding inputting to the roles of editorial staff. In most cases, little constructive consultation took place over work organization issues, although there were notable exceptions in the Croydon Advertiser (Cockburn, 1983; Smith and Quinlan, 1982) and the Daily Echo (Preece, 1987).

The hypotheses

Chapter two raised a number of important theoretical questions that this chapter has considered using empirical case study material. These involved issues of the skill content of jobs, following new technological change, the decision making process in introducing new technology, and industrial relations considerations, such as the influence of trade unions. From an analysis of the existing case study material, it is apparent that the effect of new technology on work organization, remains an unresolved issue. Similarly, the extent to which change is planned, or dealt with on an *ad hoc* basis, is unclear. These issues have been considered in previous research, but because of their contentious nature they would benefit from clarification through further research. Other areas related to new technology introduction have enjoyed less attention such as the allocation of work, trade union knowledge of, and involvement in, work reorganization, and trade union conflicts, as a result of occupational mergers, that cut across traditional job demarcations. New hypotheses have been devised to assist the exploration of these areas more thoroughly.

Writers on new technology and work organization, have often attempted to attribute new technology introduction with correspondingly specific forms of work design (Blauner, 1964). Others have not denied the existence of social and political forces, but have claimed that a key determining factor is economic criteria, and this leads to a deskilling of work (Braverman, 1974). However, existing studies do not seem to provide a general confirmation for either of these claims. In chapter five, the first hypothesis sets out to shed further light on the technology and job design controversy. This hypothesis claims, that the expansion of tasks in jobs, will be found alongside deskilled work, and there will be no general tendency in work reorganization following new technology.

Work allocation has received little attention. However, it is an issue that is inextricably linked to job definition questions. This is because the decision of who to allocate to new and modified jobs, requires consideration of the closeness of fit, of the person, to the new job. The hypothesis argues that worker allocation is not based on the most appropriate person for the job, in terms of their skills, training, and experience, but on the cost and the degree to which they present personnel problems. This may result in managers applying various social and political criteria to exclude workers apparently suitable for redeployment (Zimbalist, 1979b). Conversely, trade unions may be able to apply pressure to retain jobs that are more suitable for others to perform (Cockburn, 1983). From evidence in existing case studies the basis of worker allocation is unclear. Whether managers allocate work mainly on the basis of technical and economic criteria, or whether they place greater emphasis on socio-political criteria, is equivocal.

Existing case studies have demonstrated that planning for new technology is rare, and rationales tend to be wide ranging. Furthermore, discussion is often confined to small groups of managers. The hypothesis in chapter six that deals with these issues, mirrors the existing case study findings as expectations in the new case studies. The hypothesis is broken down into three subsidiary propositions:

(a) Planning or strategy does not exist or is limited in its coverage. Often managers respond to the need for change very much in a piecemeal manner, making changes as and when required instead of planning ahead (Rose and Jones, 1985; Child, 1972).

(b) Rationales for change are varied and do not necessarily reflect any general motive of control over workers. Rationales are varied because the lack of strategic planning leaves scope for individual managers to decide on their own objectives, which inevitably vary with personal beliefs.

(c) Decision making in advance of new technology, and in respect of the new technology reorganization is largely confined to managers. It is evident from case studies that workers are involved very little, if at all, but there are a number of other actors that can be involved. These include parent companies, employers' associations and line managers. These may all have important effects on the outcome of change. (Trade union involvement is the subject of a separate hypothesis).

In local situations, experienced and well informed trade union organization is essential. Thorough knowledge of relevant issues is important, if unions are to take an effective role in new technology discussions. Case studies suggest, that while some trade unions are able to make a solid case for inclusion in discussions, and negotiate effectively (Batstone et. al, 1987), others even fail to see the relevance of their inclusion in dialogue over issues such as choice of new technology, and forms of work organization (Wilkinson, 1983). Hypothesis four claims that trade unions will lack knowledge of job design and work organization issues, and as a result the extent of their inclusion in discussions is limited.

The tendency for new technology to affect jobs is clearly established, and the nature of part of that effect is a tendency for job boundaries to become blurred (Jones, 1988). Few researchers have considered the consequences of occupational mergers. These mergers may be due to a wish to improve the versatility of workers, or to facilitate the removal of some employees. The evidence that is apparent in existing case studies, suggests that trade unions will face conflict with each other, because they represent opposing interests where one member gains skills or tasks, at the expense of his/her colleagues.

Conflicts are evident in metalworking (Jones, 1982; Wilson and Buchanan, 1988), and they are to be expected in newspapers, because of the tradition of inter union conflicts (IRRR, 1985). In office studies, there is a possibility of problems, although none are apparent in the existing case studies presented here. The demarcation hypothesis argues that, increasingly trade unions will be forced into conflicts with each other. This will weaken the trade unions' claim for involvement in discussions, particularly on issues of work design. Furthermore, where one union represents a number of skill levels, conflicts are also possible within different levels of the same union.

In the following chapters, these hypotheses are tested against evidence collected in six case studies, reflecting similar sectors and technologies as the cases presented here.

Notes

1. See Burnes (1989) for example.
2. See also Rose and Jones (1985) who argue that things are often pragmatically conceived in an *ad hoc* fashion. See also Child (1986).
3. See Bryn Jones (1988) on flexible automation.
4. Compare with Jones's (1982) work on CNC where weak union organization also provided opportunities for managers.
5. This Marxist analysis of job redesign clearly confirms Taylor's (1947) regime of simplifying tasks. Embodied in this form of reorganization is an increasing ability of managers to establish and maintain control over workers. Since a thorough managerial understanding of what the work really involves can frustrate any false claims about the time work should take to complete.
   Although some writers argue that workers tacit skills prevent this from happening (Kusterer, 1978) this is only true to the extent that managers appreciate the importance of tacit skills.
6. See also Zimbalist (1979b) on the New York Times.
7. This suggests some evidence of attempts to humanize work, however, the use of job rotation has been seen as a covert means of gaining control over the workforce (Palloix, 1976).
8. 'Pre press' refers to functions undertaken prior to the printing of the newspaper. The functions include composing and reading.
9. Although later the move to Wapping under the ownership of News International transformed the management/union relationship.
10. 'Chapels' describe NGA local organizations.

4 The case study companies

Introduction

This chapter deals with the background detail of the companies that were selected for new case study analysis. The first section considers the organization of the company its size, products and management structure. Section two looks at the various new technologies introduced and section three at the jobs and job changes. The chapter takes a wide perspective with the intention of placing the companies, technologies and job changes in a wider context. The following chapters focus more closely on selected areas of each company which represent specific new technologies and jobs.

Zeus engineering

Zeus, employing some 5000 people, is involved in the production of diesel engines for supply to original equipment manufacturers.[1] Shifts in the market for diesels engines has meant a diversification on the part of Zeus away from the production of engines for the agricultural market.[2] Instead, the company has tended to go into alternative markets such as marine, vehicle and industrial engines, although these are also subject to cyclical trends.

The late 1970's saw the company making severe losses which, as with many other companies, has led to restructuring of the organization and an extensive rationalization programme.

Alongside these changes, there have been important shifts in the organizational structure. Originally, there was a centralized style, with central teams of operatives in industrial and production engineering, and the maintenance function. Hence, the system was centralized in both organizational and in hierarchical terms. These have increasingly become decentralized, and in some cases, central services, such as maintenance, have been reorganized in small teams dedicated to specific divisions. This now implies less control from the centre and more accountability in the immediate work area. The new form of organization uses four general managers heading different divisions. Deputizing for the general managers are four production operations managers, beneath whom there are production managers and supervisory staff. The company is thus arranged into four sectors and each sector is run as if it were a separate business. This replaces a structure of one general manager controlling the entire factory and seven production operations managers heading each sub division of the company.

Situated close to the Zeus Engines main site is the Era factory. The factory employs 375 people and is called Zeus Components reflecting the kind of work it does. The larger part of the work is handled for Zeus main factory but outside contracts are tendered for. The factory has its own production engineers and logistics staff and is independent in services and maintenance functions, but other functions, such as, wages and personnel are centralized and located at the main site.

The biggest shop is the gear shop, there is also a pipe shop and a fabrication shop. Gears and pipes are produced in both small batch and volume production. There are day and night shifts but no unmanned running.

At Zeus the majority trade union is the AEU but there are also members of the GMB and EETPU and other minority unions. The shop convener is himself an AEU member but represents all other unions on site also.

The new technologies There are dozens of VDUs in use on the shopfloor. Every foreman's desk has a VDU as well as each group of storemen. Computers follow the progress of all engines via bar code reading, and the system also involves the tracking of materials to their arrival at the point of usage, where the tracking of the engine itself takes over.

Each engine has a quality history on the screen. There is also an engine build list which will explode down every component that goes into building the engine. It is also capable of identifying storage location, price, and delivery dates. A manufacturing requirements processing system matches the availability of the materials with the dates that a finished product is required, so that the system records the number of blocks machined in the machining areas. Hence, there are many different systems on VDU that can be accessed on the shopfloor. In some case they have replaced manual systems, in other cases they are completely new installations.

There is an electronic filing and mailing system to which managers and senior foremen have access. This both reduces the paper work required and greatly increases the speed of transmission of messages. Additional technologies include an engine test complex, in which an engine is placed in test beds, the computer then puts the engine through a test cycle.

However, the main focus in the Zeus case study is of computer numerical controlled machinery (CNC). The important changes, relevant to the study, are that previously a program would be produced by production engineers from a drawing of the required component. It is a recent development for more of the planning process to be moved to the shopfloor and the operator. Here, the operator can judge if there is tool wear, make adjustments to the machine, and in some cases, write programs.

CNC machinery does not dominate the Zeus operation, there are still a great deal of conventional, manually operated machines in use. Zeus saw the advantages of CNC in its flexibility, particularly the ability to change from one component to another, which on a conventional machine can take eight hours resetting. However, managers suggested that CNC has to be matched to the application carefully if the maximum benefit is to be achieved.

The jobs Changes in jobs have come about both as a result of changes in technology and general organizational change. The tendency has been to push responsibility further down the hierarchy and to restructure the command levels of the organization so that managers at all levels now have more accountable roles. At the shopfloor level, changes are also reflected in the type of tasks performed, across the board increases in responsibility, and substantial evidence of job merging have both contributed to the reorganizing of job roles.

Grades for shopfloor workers have been reduced from a total of around twenty four grades to just four main grades. The previous organization, with large numbers of grades identified a level of pay associated with a particular operator. Chnages mean that all operators of a similar grade and pay level are classified under a single common grade, and are flexible across all operations encompassed by that grade. For example, grades D and E were machining grades and there would be 'D drillers', 'D millers' or 'D lathe operators.' The new scenario is that the classification is a 'D machinist', who can operate any 'D machine', and also undertake setting and operating in some cases.

The new grading structure involves designations of grades A to H, however the grades A to D are those in most common use. Grade A are skilled employees, and are the only apprentice trained group that Zeus employs. This includes jig and tool craftsmen, mechanical fitters, electrical tradesmen, factory service personnel, and plater welders. Those graded B are largely setter/operators, but if in assembly, they may handle documentation to ensure that building of engines and components is to the right specification, and within the time requirements. Equally, senior progress chasers are designated grade B. Workers graded C are mainly assembly test operators, and grade D's are machining operators. Below these grades workers are unskilled, for example, grade E are routine assembly workers, both grades E and F are also labouring jobs, the remaining grades of G and H tend to fall into categories of cleaners and gardeners.

Jobs have widened and supervisory responsibility pushed down to the more senior grades of worker. Inspection for example, has become largely a shopfloor responsibility rather than a specialist function, and electrical and mechanical maintenance functions have merged to some extent. However, for others the tasks are timed and still involve considerable monotony. For example, some workers in assembly have only four minutes to complete a function on the line, either adding parts to engines or making these adjustments. This means fifteen have to be completed each hour for eight hours per day. One of the senior shop stewards commented '....if you could train pigeons to do it, I think they'd have pigeons to do it sooner than human beings.' A worker can be on exactly the same job all day.

The smaller size, and more specialized business, of the Era factory compared to that of the main site, allowed a much more detailed analysis of jobs in the area of CNC.[3]

CNC operators are B grade, although there is no B rated CNC machine, and there is no one classified as a skilled operator at Zeus. The fact that grade B is only available for workers capable of performing the setting function, means that the only way the job classification can be justified is by crossing job profiles. In other words, saying that the person is capable and available to set and run other machines. Others in the section were graded D and operated radial drills, mills, and borers. There are a few other selected operators designated grade B, who are capable of setting their own machines, and are also available to be called upon to set the machines of lower grades, and generally give assistance to lower grades.

CNC operatives work to schedules, they load machines according to priorities and then work down the monthly schedule. If there are defects in the programs, the operator has to alter the program. Given problems the operator can handle seventy five per cent of amending, which is knowledge gained without formal training and tends to be learned from the production engineer.

Production engineers themselves were also called upon to produce tenders for new business outside the Zeus organization reflecting the strong business orientation of this unit. However, programming, 'proving out', and subsequent trouble shooting, remained a key part of his, and his colleagues' work.

Artemis engineering

Artemis Engineering used to be owned by the Pearson group, but was bought out by the management in December, 1986.[4] Artemis Engineering employs some 600 people, 430 are based in Middlesex which is the headquarters and the production facility, with another 100 at a site near Bristol. The Bristol site is the research and development centre dealing with design, development, stress, and electronics. A further seventy employees are based at another site, also near Bristol, a machining facility which is used as an overspill site if the main site cannot handle the work load during peaks of business.

The machine shop in Middlesex deals with seventy five per cent of Artemis products, while twenty per cent are handled by the operation in Bristol, the remaining five per cent are contracted out. The products themselves, precision hydraulic systems for aircraft, have very long lead times. It can take about one week to achieve two to three hours machining due to complex machining processes, and long transfer times of the component from one place to another during machining. Long transfer times were highlighted as a major problem of the business, an average of around five to fifteen per cent is lost in transfer times, and for more complicated components the loss could be twenty to fifty per cent.

Three main departments operate at Middlesex on the direct side, the machine shop, the assembly department, and metal treatments area. The latter involves degreasing, anodizing, chrome plating, and hardening. The areas studied, assembly and machining, are staffed by apprenticed trained skilled workers. The metal treatment division of the factory was staffed by semi-skilled and unskilled workers and had no use for new technology.

The plant at Artemis Engineering is organized in four factories, each has a manager and is served by a central quality assurance team. Within each factory workers are organized in work groups and incorporate supervisors, technicians, and dedicated inspectors. This form of organization is so arranged to accommodate a system forming families of components, cells of machines and the workers, to avoid loss of transfer time due to bad organization of machining sequences. Work pieces are retained in areas for the maximum time possible to avoid costly transfer, and to achieve optimum utilization of equipment.

The Assembly department is split into two departments: 'new build' and 'repair and overhaul.' The repair and overhaul division is looked after by the repair and overhaul manager, and the new build area by an assembly superintendent. The repair and overhaul department exists to provide after sales service and repair which is offered to customers. This can involve 'deep overhaul' of components to include stripping down, surveying and inspecting the part, and to refurbish and rebuild. The department has thirty two assembly technicians, two foremen and two charge hands (really shift controllers). Figures do not include five assembly technicians currently being trained in the company's 'training school' (a 'Porta Cabin' outside the repair and overhaul managers office).

The New Build Assembly section also comes under the management of the repair and overhaul manager but, like the machine shop, has its own superintendent who has the day to day responsibility for the department. New Build Assembly tests and assembles all forms of aircraft equipment and flight controls. In the department there are forty two technicians plus one foreman. Most are concentrated in grade two and this denotes that they have a good knowledge of the products. There are a few grade three's, a designation that denotes they are 'supervisory material' and are usually very experienced workers since it takes a long time to acquire thorough product knowledge.

The main trade union for the manual workforce and the holder of negotiating rights is the AEU. There are five shop stewards spread throughout three main areas of representation: the machine shop, assembly, and treatment areas. Representation covers all grades of workers from skilled through to unskilled.

The new technologies 1983/84 marked the major introduction of microelectronic technology in Artemis. The most significant technological addition to the factory in Middlesex, is what management described as a partial FMS system, attained without a Department of Trade and Industry (DTI) grant.[5]

At about the same time second generation CNC machines arrived (then top of the range), and training took place inside and outside the firm. However, despite this minor revolution in the firm, some fifty per cent of components are still produced on conventional machines. This, nevertheless, is a significant decrease on the eighty per cent that were being conventionally machined five years ago. The machine shop superintendent expressed a strong preference for CNC machines based on their greater reliability even though they require regular calibration checks.

Part of the purpose behind the introduction of CNC, was the need to reduce the number of operations that a component had to be put through. This in turn was to reduce lead times, and has been achieved by compressing a series of operations into individual CNC machines. Some machines are now capable of performing one operation which would have meant five separate operations on a conventional machine. Equally attractive was the fact that CNC machines produce in the region of two and a half times what a conventional machine can produce. Therefore, less than half as many people are required to run them. Also unmanned running will reduce the need for as many people, but maintain or increase production hours. In these ways new technology is used to alleviate the problem of the skills shortage that Artemis complained of.

In the repair and overhaul department, computer test equipment is employed mainly in fly-by-wire assemblies. This takes the form of a computer program which lasts for one and a half hours, and puts the system through a range of tests, reporting any under performance. In the same area, a new computerized database system replaces an old manual system for purposes of production control and to keep track of production units, this may be used in future to fault find and establish defect trends.

There is a computerized pre-test preparation used for one of the main production units. A 'burn-in test' settles the unit down and shakes it (as would occur in normal use) prior to testing. This is a completely new addition and does not replace any previous simulation, but the test that follows this preparatory operation is significantly quicker than testing was without the preparation.

The jobs Overall there is a new flexible approach to work in the sense that workers are required to be more flexible across a range of functions. Also more responsibility is taken by individual workers for quality assurance. In general terms the increased use of new technology has meant a shift away from traditional machining skills.

Job gradings have also been changed, and there can be little doubt that this was largely to meet the changing skill requirements. In the machining areas the grading structure operates along the following lines. Technician[6] grade one is an apprentice trained in grinding, turning and milling, but is confined to the operation of conventional machines. Technician two is trained on CNC, but also has a background of apprentice training on CNC machinery and displays greater awareness of quality control. Technician grade three is involved in planning and programming. Here there is an overlap of functions, at least in theory, with production engineers. Clearly the grading structure in the machine shop closely reflects the technological developments in that area and particularly in CNC.

Artemis makes use of unmanned running and multi-manning. An important part of an operator's job at the start of a shift is to organize the work so one machine can be set running while another (possibly the machine that has run unmanned overnight) can be unloaded. Operators move between machines loading and unloading in between production cycles. Production engineers, reportedly, would often work closely with CNC operators through the 'proving out' stages, and subsequently in fault finding.

The workforce in machining is 'balanced' between CNC and conventional skills, because, as the manufacturing manager maintained, there will always be a need for conventional skills and knowledge. In keeping with this is the retraining and educating of existing workers, rather than adopting an approach that would involve staff not familiar with conventional machinery and would confine training to programming knowledge. The importance of this philosophy was confirmed by the senior shop steward who argued that it is important to have background skills in conventional machining whether working CNC or not. This is because working on CNC's involved much more responsibility and the experience of conventional machines gave the operator vital knowledge.

Before the changes, the firm employed setters and operators as distinct groups so there were a number of semi-skilled operators. With the changes, the workforce became composed of skilled setter/operators, and the semi-skilled were lost either by natural wastage, redundancy, or early retirement.

The same grading designations as the machine shop are followed in the repair and overhaul department and in new build assembly. It is not uncommon to interchange labour between the two departments according to need, a demonstration of labour flexibility in the plant.

Summary - metalworking studies

Both Zeus and Artemis, although different in their size and the products they produce, display similarities in their approach to a decentralized organization. For Zeus this is largely a response to the crisis of over manning and is part of a rationalization programme. Artemis adopts such a form of organization in order to reduce the long lead times and down times that are a characteristic of the products the company produces.

Zeus displays a broad use of new technology in many areas of their plant. The size of the organization, and nature of products made their use of advanced stock control systems, electronic mail, and detailed production control systems, viable propositions. Artemis, on the other hand, have little new technology usage by comparison, due to their smaller size and smaller financial resources. However, similarities exist between the companies in their use of CNC machinery, and despite the different skills employed in each organization, there are similarities in the work that CNC operators are required to do.

In both Zeus and Artemis general changes to work organization have taken place spurred by new technology changes, and other factors such as the need to rationalize. In Zeus work flexibility was required to allow the use of fewer staff to save costs, in Artemis the shortage of skills was an important factor in changing jobs and initiating the internal training programme.

Zeus particularly, have a number of technology related changes to jobs. Maintenance workers are interchanging skills as a result of the increasing complexity and blurring between mechanical and electrical trades particularly evident in CNC machinery. Like Zeus, Artemis have redefined the role of their CNC operators to include aspects of planning, and have reorganized the grading structures to recognize the additional work that is taking place. In both companies self inspection has replaced a central team of specialist inspectors, although there does not seem to be any obvious link with new technology here. It is, nonetheless, typical of the kinds of changes that are taking place.

New technology and industrial relations in the newspaper companies

Available since the 1960's, photocomposition technology was held back in many newspapers by skilled traditional workers who had effective control over the work by virtue of the exclusive nature of their skills and a cohesive trade union, the NGA.

For many years printers unified by the NGA were able to control much of what went on in the national and provincial press. Although this was particularly true of the national press, the trade unions still had significant control in the provinces.[7] New technology which would simplify the jobs so that, allegedly, anyone with basic typing skills could do them, was held back by the unions in an attempt to retain their skills and their jobs.

Tension built up as a result of this resistance, management feeling disturbed about the long domination of the NGA, which they wanted to sweep away. However, the NGA eventually began recognizing, and reluctantly accepted, the inevitability of new technology introduction. To minimize the impact on the industry and the NGA members, the union proposed terms to introduce new technology in such a way that NGA operatives would be retained by the firms. Hence, the NGA began to think in terms of working with new technology.

The new technology options were the use of a photocomposition system which would replace the old linotype machinery with computers. But the more controversial side of the new technology introduction was to be the direct entry of text by newspaper personnel, such as editorial and advertising staff. This would drastically reduce the need to maintain a separate inputting section, which was the role that a large proportion of NGA staff had always performed. Direct entry of text was an aspect of the changes that the Newspaper Society (NS), the employers' association for the regional press, emphasized. They argued that newspapers needed this kind of technology to survive, and highlighted the growth of free newspapers, and the consequent loss of advertising revenue, reductions in circulation figures, and the closure of newspapers around the country (Newspaper Society - Project Breakthrough, undated). While the futility of typing the work twice (the double keystroke) was the central focus of the NS the cost considerations associated with this were not the only ones.

The NGA objected to the way in which the Newspaper Society were going about their Project Breakthrough campaign for the wholesale, universal and immediate acceptance of new technology. However, the NGA was recognizing the inevitability of new technological change (Wade, 1982), and Dubbins (1982) proposed a progressive integration of direct entry technology.

The newspaper society's Project Breakthrough failed in its goal of achieving a national enabling agreement for new technology introduction with the unions. But eventually employers began the introduction of new technology and achieved negotiated change with the NUJ and the NGA. There was considerable inter union conflict,[8] and the NGA published a document outlining an approach to new technology introduction that they would like employers to adopt. This was entitled *The Way Forward: New Technology in the Provincial Newspaper Industry* (NGA, 1984). The employers rejected this as did SOGAT, since it aimed to preserve as much NGA membership as possible, even in cases where NGA staff were re-employed in traditional SOGAT or NUJ areas.

Finally, the NUJ and NGA struck a deal in 'the Accord' which established that they would negotiate jointly, and decide the course of action together. It also resolved disagreements over which union would represent NGA workers redeployed in the editorial area.

It was against this background that the negotiations for direct entry took place in each of the following case studies.

Mercury and Hermes newspapers

Mercury newspapers, a subsidiary of the large newspaper group Westminster Press,[9] employs some 270 people including thirty five part time. The company was reorganized in 1970, and the profit margin has since leapt dramatically. This is not due entirely to the reorganization of the company, but must also be seen in relation to the growth of the town in which the newspaper is based. Three main newspapers are produced. The first of these is a daily evening newspaper and of the two other papers, one is produced weekly and the other, also a weekly newspaper is delivered free.

Hermes Newspapers is a very old newspaper company established in the late 18th century. Previously a family firm, in more recent years it was also acquired by the Westminster Press Group with the agreement that the managing director would remain. At the time of the interviews there were 230 employees working in the organization. The newspaper produces a variety of titles including the main newspaper which is a daily evening tabloid, three weekly newspapers and one weekly 'free sheet.' The company does not enjoy the circulation of their counterparts at Mercury Newspapers.

Both companies have three main divisions. There is a production department, advertising department and editorial department.

The production department breaks down into 'pre-press' and 'press.' The press area looks after the printing, distribution, and circulation of the newspaper. Pre-press involves three broad areas: composing, paste up, and plate making where the final page make up is photographed. A printing plate is made from the negative which goes to the press.

Much of the distinction between staff in the newspaper industry is owed to tradition, not least is the split within production that in the past has always divided traditional skilled workers who are NGA members and SOGAT who are semi-skilled. Composing, negative and plate making, and press management are skilled NGA areas. The labouring side of the press and despatch are SOGAT areas, and at Mercury Newspapers it was reported that some of the staff employed in advertisement telephone sales ('tele ad' staff) were also SOGAT members.

A further split in the workforce is dictated by the nature of work, but is further emphasized by the trade union organization of the National Union of Journalists, and the National Graphical Association. The production areas were closed shops, and the majority of journalists were NUJ members at the time of the study. The Institute of Journalists (IOJ) also existed at the newspapers but with very few members.

The new technologies The changes in both newspapers are quite typical of the industry's general transformation. This is the shift from a traditional system of mechanical operation to one that, in the first stage, involves the use of computers but retains the old workforce. In the second stage much of the production workforce are replaced in concert with further technological changes, to facilitate the direct entry of text. This easily discernable step by step approach of implementing the new technologies is also common to other newspapers.

The old linotype machines, previously used, had non standard keyboards and were designed so that they would deliver a letter or character, made in metal, down a chute into a galley (collecting tray) which would eventually form a page. When the tray was full it would be covered in ink and a paper copy would be taken, which was known as a 'galley proof.' This would be read for corrections by NGA readers.

Once amendments had been made, the stereo department embedded the galley plate into high quality papier mache, which was then dried into a curved shape. Molten lead was poured over the mould to make a curved stereo plate, which would fit on the press. The plates could weigh fifty four pounds, and they would be carried by men to the press. Today plates for the Web offset press weigh about one pound. The work was skilled and there was a strong element of pride in the job, since the compositor was a key operative in the organization, although mechanics were on hand to assist. There was a clear masculine association with the dirty and smelly environment in which they worked. Additionally, there were health dangers from the lead oxide deposits.[10]

It was not until 1981 at Mercury, and 1984 at Hermes, that the newspapers left behind linotype production, and went to photocomposition systems. The next phase, following replacement photocomposition systems. The next phase, following replacement of printing presses, took place in Mercury from August, 1987 to Spring, 1988. For Hermes, however, the process started in August, 1987 and took a full year to complete. This phase was to prepare the ground for the introduction of a 'direct entry' system, to enable all staff in both the editorial and advertising departments to key their own work. This change was coordinated by the deputy editor at Mercury Newspapers.

During the period of negotiation at Hermes the project coordinator changed, and a new managing director was appointed. The first project coordinator was the production manager. After the managing director had left and a new one had been appointed in December 1988, the deputy editor became the new project coordinator.[11]

There seem to have been two planned phases of new technology introduction which can be identified as 'production take photocomposition' in which the radical change is the technology. The jobs change but the workers remain the same. The second phase of direct entry photocomposition actually changes little in terms of technology but has a major effect on the jobs and organization of the firm, since there is a change in the way the organizations are staffed. Hence there is an important difference between production take photocomposition systems. What actually changes is the copy flow structure[12] which remained the same in production take. However, in direct entry, the copy arrives via the computer, linked directly to editorial and advertising staff.

Around the offices there are a range of linked VDUs which can be accessed by journalists and advertising staff. Each story is assigned to a basket or a computer file. There are a range of baskets for different specializations, for example sport, news and features. The final decision of where the story should go is for the chief sub editor and stories are passed on to the next stage by means of a computerized queuing system. So a story is written by a reporter and then sent to a news editor's queue who alters it, and sends it to the chief sub editor's queue. The chief sub editor farms out the work to other sub editors and then it is passed to a page basket ready for the works to call it up. The page is then made by photographing and then pasting up the bromide (the photographic image, rather like a negative) in the same way that it has been done since the introduction of photocomposition. However, this is likely to be short lived and once electronic assembly is established the process of bromide pasting between the page production and the printing press will disappear.

The jobs New technology systems have made it possible to amalgamate three traditional groups of production worker, press operators and despatch people and those who make plates. It has been possible to merge the tasks of this group so that all participate in each others jobs. Previously, there have always been strict demarcation lines laid down by the union preventing semi-skilled SOGAT workers from doing skilled NGA work.

The job changes are common throughout the industry and are well known, the old method of working was on large heavy mechanical machines. These machines produced metal type in lead '....it was heavy and it was dirty and it was really a masculine job' (Production Manager, Mercury Newspapers)[13]

The initial change in the newspapers, that of hot metal to production take photocomposition, failed to have a very great impact on who was employed in the jobs, or in general approaches or work attitudes. But the jobs changed quite fundamentally for NGA staff, and compositors in particular. Working conditions became cleaner, and less manual work was required. Now the compositors set type on more modern visual display terminals rather than mechanical devices and were dealing with paper rather than hot metal. The environment changed, and the operation was transformed from what was described by managers as light engineering, to the ambience of what was almost an office. There were some redundancies as a result of the new system, but these were few in number and purely voluntary. Retraining of compositors was carried out since the new keyboards were different from those on the linotype machines, but for the journalists, changes were virtually non existent.

The segregation of production and editorial jobs, and between skilled and unskilled production jobs had been clear in the past. Journalists were doing a literary job in reporting and editing. The reporter would type a story, which would then be corrected by the sub editor and passed, with layout instructions, to compositors who would compose the work. Composing involved the preparation of text with the machinery available.

The involvement of the semi-skilled SOGAT members would occur in publishing, in the press room, and in distribution of the newspaper. This was conventionally, the only point where they were involved. New technology has facilitated the destruction of such demarcations and the compositors have been replaced with non union staff handling the basic function of inputting contributed copy.

At Mercury Newspapers, the new staff have been called editorial and production assistants and a policy of employing part time, short term contract and highly flexible labour (usually female) has been pursued. Pay levels for these inputers have been established well below that of the compositors.

Due to the change in methods fewer of the traditional production staff were required. However, some of the remaining composing room staff deal with the 'contributed copy' which may be local sports reports, feature material from freelance journalists and readers' letters. Therefore, the new division of labour means, that the reporters are taking the inputting role from the compositors, and the sub editors are adopting the role of organizing the work and arranging typefaces.

Also in the 'tele ads' area the procedure used to be to fill all the details out on to a form which is then passed to the compositors for setting. With direct entry workers are typing the details straight onto the screen and into the system. However, this only applies to small classified advertisements. The display advertisements still go to the composing room and traditional workers. Nevertheless, the direct entry system in advertising resulted in the loss of six jobs. Unlike Mercury Newspapers, Hermes decided not to retain specialized readers to proof read advertisements copy for errors, although the production manager suggested that this would be under review. Bonus payments were also paid to some outgoing compositors in return for assistance in handling the new technology change over.

Mercury management wanted very much get away from the excessive rates of pay that compositors were receiving. Journalists with a university degree and four to five years service were earning forty to seventy per cent less than a compositor, largely due to the control exercised by the union. But the company has achieved an equality between those who are journalists and those who are undertaking similar responsibilities in advertising.

Hence, there are significant effects on jobs held by skilled production staff and important repercussions for the NGA chapel. It is interesting to note that more staff are being used to take on the menial side of a journalist's job.

Summary - newspaper studies

The two newspapers are very similar although Mercury is the larger and more successful than Hermes, and displays a different style of management. These companies need to be considered in the general historical and political context of newspapers. This is very important in understanding the way new technology was implemented, the significance of union opposition, managerial approaches to new technology, and the way jobs were reorganized.

In the newspaper companies the new technologies employed are almost identical. Even the pattern of implementation is similar. In both cases a two stage approach to technological change is discernable. The first to production take photocomposition, and the second to a direct entry system. These changes in new technology not only mean a transformation in the management of the business, but also mean radical changes in work organization.

The newspapers have identical new technologies. The only difference between the companies being the time scale of introduction. In both cases similar patterns of new technology introduction are followed. The first stage, from linotype composing to a production take photocomposition system, involves a change in the nature of the work but retains compositors for the composing function. In other words, the tools and jobs change but the staff do not. It is clear that the work is simplified and no longer requires the specialized knowledge that it did before.

The second change means that journalists begin to adopt the inputting function and where this is not possible a new lower paid, and less skilled group of staff are introduced into the companies to provide the inputting function. The reporters have more responsibility for getting their work right and sub editors become the last line of checking. In the case of Mercury some readers have been retained to check advertisement copy for mistakes to avoid loss of revenue through errors.

Apollo aerospace

The firm Apollo Aerospace, is a subsidiary of a large diversified organization. The business is centred on the production of aircraft systems such as computerized 'head-up displays.' These are very costly as a result of the development work that goes into them. The production of such systems is very labour intensive and the firm employs highly skilled labour.

The main site, houses clerical, manufacturing, technical and administrative staff. The site overall has increased in numbers since 1980 to 3200 but since then the clerical strength has decreased by fifteen per cent, clerical staff now number about 400.

Two divisions of the company were subject to detailed analysis, the accounts department and the inventory control area of the Manufacturing Services Division. The responsibility for the accounts department is divided up between two accountants. There is also a systems accountant who deals specifically with computer system development and the trouble shooting of existing systems. In total there are twenty five people in the department. The payroll department has nine staff, there is a payroll of 3200 people of whom forty per cent are paid cash every week. There are three payroll areas: management level monthly paid, standard weekly and hourly, and shift workers. Since, legislation requires records to be held for sickness in the statutory sick pay scheme, greater computer power is necessary to contain this.

The Manufacturing Services Group (MSG) is a division of Apollo and operated like a separate business, although MSG is not a profit making concern. Its function is the supply of other groups in the company. The MSG take orders from different groups, and supply at factory transfer price or cost price. There are 550 employees including direct and indirect employees, the group is strongly manually orientated and has forty per cent of all the direct workers on site.

The group is split into four sections.[14] The key managers in the department are an operations manager and industrial engineering manager. A production manager looks after the sections, and a materials manager supplies a service to them in material sub contract purchasing. There is a chief accountant and budget and administration manager, but most of the operatives report to the production manager. There is also Central Services

Group which comprises of goods inwards and site services sections.

Trade unionism at the company varies according to employee status. Clerical staff are members of APEX and MSF represents technical and supervisory staff. The main union in the plant production area is AEU but there is also a GMB and TGWU presence. The manual unions are treated as one group, but as the majority trade union the AEU retains the negotiating rights.

The new technologies Amongst the managers, there were mixed perceptions of the company's level of new technology development in the office functions. However, it was noted that on the manufacturing side of the business, the number of computers, and computer applications had increased tremendously. There is considerable use of computer packages involving complicated software in technical functions. The staff have expanded on the software side as a result. The manufacturing side had also been using advanced CNC machines for sometime. This seemed to be in contrast with the relatively poorly developed, or recently developed, office support new technologies. This is true, even in office areas that are directly involved with shopfloor operations, such as production and inventory control.

Production and inventory control is now managed with the use of a manufacturing management system. This involves resources planning, work in progress control, accounts data, and standard costings. It also holds some purchasing information. In this way it bridges the gap between the shopfloor and the office but is mainly office controlled. The system will advise an enquirer as soon an item is withdrawn from stores. It will also advise that a new item is coming into stock, and will arrange an order to be raised if this is necessary. In addition it can estimate call off rates because the customer demand will have been noted.

There was previously a manual system of stock checking in which eight people were employed. They had to visit a location and record everything that was there. Now the computer records everything that is in a particular location, and is able to give a detailed breakdown of the part number, factory transfer price, date of the last stock check and so on. The number of people required is reduced to only two.

Considerable advances have been made since the first accounts package was introduced in 1979/80. The original system was a batch mode system (data had to be prepared in batches before passing to the mainframe) and had considerable limitations. Providing the 'posting' of the forms had been finished by midday, processing would take place overnight, and the printout wouldn't come until the following day.

The new system is an integrated accounts package, and was introduced in August, 1986. It is capable of giving printouts whenever required. The new system processes and prints in twenty minutes after finishing 'posting.' Originally there were two terminals running the system, but now all but one person has a computer on his/her desk. In fact there is access available for up to seventy users.

The payroll system, separate from the general accounts system, is capable of handling any standard deductions such as union dues and pension contributions. Although supplied as a standardized package, considerable modification was necessary to suit the company's requirements. However, the bulk of the system is based on legislation, which is common to all firms.

The well documented changes, in the transition between typewriters and word processing, were described by the union secretary of APEX, who herself was the supervisor of a wordprocessing section. In typing pools documents were typed from engineers' hand written script, but the engineer would want to change it the next day and the typist would then re-type the article. However, now engineers have begun to type their own work on word processors and save it on disk, it is then reworked by typists with little extra effort.

Union representatives also pointed out, that because information is controlled within departments, the need to have people devoted to inputting data has receded significantly and data processing has all but vanished.

The jobs The company's grading system has three groups of manual, clerical, and professional and engineering. There are distinctions made between direct and indirect staff, in terms of hours (40 and 37.5 hour weeks respectively) and other benefits. However, the operations manager of the manufacturing services group felt that this distinction should be removed. All directs, and a few indirects, 'clock in.' Skilled manual workers come under the indirect workers designation.

In the production control department of MSG, the manufacturing management system project was started in 1981. Since 1983 there has been a five year period in which directs have been reduced by around eleven per cent, but there has been an increase in the number of indirects. For example, the raw material production department used to consist of six staff, but now has only one because the computer that receives the information about raw materials, saves the work of manually writing out the necessary details. Also because the standard of information is so much higher, the number of progress chasers required has reduced significantly.

Similarly, for the process clerks the workload has reduced and what now takes one hour, could have taken three days without the computer. The previous scenario was heavily labour intensive, and involved the updating of white master cards (documents being the main information source relating to orders passing through the factory) with fresh information about the status of an order. This would be done by removing the white master from a file, and updating with tickets that had been received from the shopfloor, and then replacing the white master. There was a constant process of pulling cards out and putting them away again.

The progress chaser now works on a VDU but previously would have had to go to the cards, lean over the staff responsible for updating (all female), and then find the card was not there but was being dealt with elsewhere. Invariably the progress chaser would give up. The problem was so great that there would have to be times when no one was allowed near the cards to enable the staff to update them. There are now two women in this area, whereas prior to the change there were nine women working there. Hence there are major changes in jobs, and these have meant the learning of new skills in areas such as accessing a VDU.

In the Payroll department new technology has allowed the absorption of extra work and offered a staff saving of one. Work on the monthly payroll, previously handled by outside sources, has now come inside the firm and the information is run on house computers with external software. This has given much greater control and allows the department to do automated costing, making the flow of information more effective. Union due collection and pension deduction is now done in three payrolls, not two as was the case before. The supervisor suggested that much of the more boring side of the work has gone and the individual is left with more interesting tasks.

The office is split up so that one part deals with staff payroll (weekly) involving two people, the other end of the office deals with monthly and hourly pay. The hourly payroll is largest because it is on clock cards. Clock cards are received Thursday afternoon (or Friday morning if overtime is involved) so most of Thursday afternoon and Friday morning is spent working on cards. Aspects such as hours, overtime, and holidays are calculated and input. The sick list is also checked and the system is set to calculate statutory sick pay according to legislation.

The computer calculates the hours and times to get an hourly rate and also, calculates national insurance and union dues, all previously manual procedures. On Mondays the Payroll is produced which is in the form of a bank tape, prepared for issue to the banks for payment through personal accounts. The tape is checked and amendments made for large errors but if there are only small amounts, adjustments will be made the following week. Reports come out each week such as pension contributions, and these have to be manually checked.

The average grade is clerical grade two, but people come in at three or four and if they meet the requirements they are promoted to the higher grade.

The invoice clearance office pays all the invoices that come on to the site on behalf of the operations groups. There are nine staff, three do the same thing but have blocks of the alphabet relating to different suppliers. There are some 3000 suppliers and regularity of contact varies with the size of the items supplied. There is also a VAT clerk, three bought ledger clerks and one foreign payments clerk.

Bought ledger clerks work with a large alpha/numerical filing system where 'live orders' are held, these are matched with goods received notes and checked against the advice note from the company. If there are any discrepancies they are passed to the purchasing manager who decides whether to raise discrepancy notes against the customer. Each supplier has a unique bought ledger code and the code will give access to all the current information held on computer, such as what is to be paid and the bank details.

Prior to this the details of a new supplier would have to be sent to London and they would send back the details which would then be recorded on the invoice and passed for payment in batches. The information would be returned to London where payment was organized. Under this system tracking down invoices and establishing what stage had been reached with them was very difficult. The mainframe computer in use in London was inaccessible from the local site.

The VAT clerk's job involves the checking of invoices for a valid order number and then highlighting the invoice number, invoice date and the VAT to help the punch clerk, who is responsible for putting all the bought ledger codes on each invoice, and selecting the document type. This information is entered on to the screen and each document is given a unique document number, so that document number and bought ledger codes are cross referenced. The computer system is also cross referenced by company names.

Despite the use of computers there are clearly a number of manual functions that have not been lost in the new technology transfer.

Achilles shoes

Achilles Shoes is the largest supplier of shoes in the UK and at the time of the study was still a private company retaining strong family control. In addition to its manufacturing and distribution operation there are around 800 shops controlled by Achilles including sub groups under other brand names. The group has 500 wholly owned outlets and 17-18000 customers which range from small shops to department stores. There are something like 200,000 lines of orders received every day, and the filling of the orders is further complicated by the sixteen sizes of shoes available, as well as the different fittings and varying colours and styles. There are 10,000 staff employed by the group of which 4-5000 are employed in Achilles owned shoe shops around the country. It was the accounts division of the business and the distribution and customer services division that were selected for study since they had gone through, or were about to go through, considerable new technological transformation and development.[15]

The distribution and customer division of the company has two warehouses, one dealing with Achilles brand footwear, the other with specialist parts of Achilles branding. There is a clear separation between clerical and administrative employees, manual warehouse, and technical occupations.

The accounts division comprises a credit manager who looks after the credit control operation and sales ledger. The financial accountant heads the department and is responsible for bought ledger, general ledger and payroll as well as the provision of financial statistics to shops. He also looks after the development of computer software in the department. The UK and overseas previously had separate sales accounts sections and these have since been amalgamated under the credit manager, although in Europe some credit control is localized.

The only clerical union is APEX and there are about 350 members in the branch. MSF are the union looking after monthly paid staff, but there are many other unions on site and elsewhere in the group including the EETPU and the TGWU. The main unions are NUFLAT, MSF and APEX.

The new technologies Achilles had already made considerable advances into the use of computer technologies, but were further developing systems at the time of the visits. Electronic point of sale (EPoS) technology was being installed in the wholly owned outlets. Here the computer records sales during the day, and sends back the information to the host central computer at night so that it is possible to gauge accurately what has been sold nationally in a day.

The personnel department had undergone some minor technological changes, and was using a computer system for the retention of information and to access information using a variety of cross references. Information is kept about employees by department and division, and is cross referenced by age, grade, sex and pay status (hourly pay or monthly). This greatly assists planning and administration and the information can be transferred easily to payroll or other relevant areas. Passwords allow other departments access to the system to see some personnel records, but only those relating to that department. Also the wordprocessing facilities in the department have reduced the routine letter writing considerably.

Every year staff have to be informed of pay increases and this can now be done addressing letters to employees personally. Other savings include less time spent in retyping routine letters and in filing. The personnel manager suggested that new technology systems were not used greatly, but for personnel and other departments people now have '.....more accurate, more complete, more up to date and less subjective information available to them.'

In 1980 a team was set up to look into the development of the warehousing facility which involved, not only new equipment, but also new working practices. However, the introduction of computer systems were at the core of the changes. Mainframe applications already existed, but with the introduction of a micro computer, a work in progress (WIP) system was developed. The warehouse was already heavily mechanized, but the idea was to use the computer to solve logistics problems rapidly. It was also intended to eliminate duplicated work and to reduce the requirement for heavy manual work. The WIP system helps office staff in planning more efficiently, and will give advice to shopfloor workers. In addition, records are kept of how much work should be put through the warehouse and where the work is. Work passing from one part of the factory to another is registered when items are subject to bar code reading. There is a consumer liaison and retail link at head office, which can access the warehouse system and find out where a customer's order is, and what stage it is at.

In addition warehouse technology was installed to enable the selection of shoes for 100 customers at one time. Manual 'picking' of shoes to individual orders was a highly labour intensive process and hence very costly. Shoes picked in bathes are sent down a conveyor, each labelled box is laser scanned, and from the information received, the computer will direct the shoes down different routes according to customer. In this way the computer advises and selects for each customer, and is capable of pointing out errors in picking. The main advantages are considerable savings in labour and more effective utilization of the remaining labour. The system will also collect customer account information by keeping records of the products and where they have been sent.

The accounts department had been using a punch card system for twenty years, and at the time of the visits was still using this system. However, an integrated computer was to be introduced into the department. Some use had already been made of localized computer facilities, the overseas sales ledger and the purchase ledger was already run on an independent computer.

A payroll package was introduced when the other system became outdated. But the sales ledger, a major part of the work load in the department, was using punch cards fed through the mainframe computer. The new system is required to advise what is owed by a particular customer. To do this on the previous system involved going through copies of statements and adding them together, which was a very time consuming process. The system is also expected to handle reporting of overdue accounts automatically, and to provide reminders to this effect for customers. The system will act as a general information source for general accounts data, and provide more specific data on individual customers.

The manufacturing side of the business is also involved in updating the technology it uses. CAD/CAM is now employed for the design of shoes and this provides an example of new technology removing skills to the office and away from the factory.

The jobs   According to the systems manager at Achilles, new technology computer systems have meant that the employees 'can give the job more time and more attention.' He also argued that there was significant reskilling involved suggesting '....they have to drop one skill and learn another....'

Management at Achilles were anxious to play down the role of technology in its effect on jobs, and commented that no jobs were lost as a result of new technology. Although some people had retired early as a result of the introduction of the computer, the motive was not to introduce the computer to actually remove staff. In other circumstances some staff have not been replaced as they have left. But the message was clearly that the transition to the computer was quite unproblematic, and although it has had important effects it was undramatic.

However, there is ample evidence that, in many parts of the organization, new technology has meant a radical transformation from the old style of working. For example, in the case of EPOS, involvement in paperwork is reducing considerably, some information is captured automatically in the till and some is transferred to the use of a keyboard. The effect is mainly on the staff in the shops, but, in the longer term, it is possible that the head office jobs of those involved in vetting paperwork from shops will reduce.

In the warehouse the distribution and customer manager suggested that '....the effect on staff and management is more involvement in development, more awareness of what computers can do for you, and more use of computers, particularly the local computers to organize their day to day work.

In distribution all grades are involved in some sort of computer operation, although not all are trained to use terminals. Those involved in bar code reading, or entering details of the changed status of goods into the computer, are one grade lower than those using the system for enquiry purposes.

Clerical staff are scattered across the factory in the locations of the equipment that they have to run, and as a result of the computer systems, there are significantly fewer people working in clerical and administrative jobs. Previously jobs would involve much sorting of paper for administration and processing. There were a team of sixteen women who used to prepare orders for the warehouse, now there are three because the sorting, ordering, bundling and numbering is done by computer.

However, there are two computer operating jobs which never existed before, and they deal with the claims paperwork that was handled by a team of people before the changes. As a result these tasks are much more organized and more information is available. New jobs are slightly higher grades than previously was the case. Typically, they are one grade higher than they were before the new technology changes, and this is in recognition of the new tasks that have to be performed.

Clerical staff have reduced from a 1974 level of thirty seven to a 1988 level of just eighteen. The distribution and customer manager explained that jobs have changed, but the work is still tedious, and gradings have not changed. Other areas of change are consigning, which has reduced from ten to three people, because bar code scanning organizes a consignment for a vehicle. Before this all consigning information had to be punched into the computer for every load. Also claims clerks have reduced by one but they do the same work and there is considerable room for improvement in computer usage, although it is not yet a priority.

On the manual side there have been significant changes, some forty four employed in picking in 1974 have reduced to just nineteen in 1988. Before the computer took over the control of bulk picking, there was an interim stage when people were picking in bulk and then manually sorting boxes afterwards. Now instead of being pickers they are using the sortation equipment to move the packages. The savings stem from the reduction in lost time travelling between locations in the warehouse. More time used to be spent walking than actually picking because the warehouse is very large. Similarly, there used to be considerable room for error, and the people who used to pick shoes could easily make mistakes by picking too few, too many, or the wrong ones. There was previously a need for checkers, now the scanner does the checking and the computer verifies the order. Supervision and control are much easier as a result.

In the accounts section there are a number of clerical workers organizing and operating the firm's credit control function. Their job is to ensure that statements are sent out promptly and accurately, and payments received on the due date. There are discount terms, that need to be administered, to encourage early payment, and due dates after which the staff must contact the customer. Late payment advice letters that used to be sent are now used less and less, and much of the work is done by telephone. There is still a requirement for letters and these are handled on wordprocessors since they are largely standardized. However, even special letters are dictated and then typed on wordprocessors since they are easily amended in this way.

Greater computer usage was about to be introduced into the department. This would mean that the filling in of documentation manually would all but vanish, and be replaced with work involving the typing of the relevant data direct onto a computer system. However, there will be additions in skill since the staff will be required to decide on a customers credit rating (in terms of the number of days the customer will be allowed to pay). There are five clerical gradings, grades one to five. The credit control staff fall into grade five, the top grading, there is also a staff supervisor who reports to the credit manager.

There are two sections, one dealing with UK accounts, and the other with overseas accounts which have seven and three staff respectively, on grades three and four plus one supervisor on clerical grade five. Invoicing overseas and UK is currently handled on the mainframe and the department checks its accuracy. There is also a section that deals with invoicing for other services and products provided by Achilles. This is entirely a manual system of batch processing where the invoices are keyed on to tape and fed to the UK statement on the mainframe computer. This will become a self sufficient area with the advent of integrated computer systems.

The notion of introducing new technology had caused some problems amongst the staff in accounts, some concern was expressed at first because there is an association of computerization with job loss. However, the department has followed a policy of reducing the staff in previous years while increasing word processor usage, with the objective of minimizing the impact on staff levels in future. While the managers boasted considerable forward thinking about manning levels to avoid a radical reshuffling of jobs, they had largely failed to discuss the technology and changes with staff. Managers tended to interpret the 'impact on jobs' as job loss rather than job change and because few redundancies have occurred in the department and few were planned, little attention was being paid to the job issues.

Summary - clerical studies

Both companies in the remaining manufacturing organizations are very large. However, their products are very different and the organization of the companies is also dictated by the products they produce. Achilles has a chain of retail outlets and is a very large volume supplier with a distribution operation based at head office. Apollo on the other hand is not a large volume supplier and deals with far more complex specialized products. They remain useful for comparison since their approach to the use of new technology has been piecemeal and quite slow with still considerable room for improvement.

In the remaining manufacturing companies new technology again varies in accordance with the products produced. However, evidence of CAD/CAM at Achilles in shoe design and DNC at Apollo demonstrates that the manufacturing side of the businesses are adopting new technologies and may be rather more developed in these areas than the office functions. The new office technology is certainly not so prominent in the business. Its introduction varies in extent and priority, and the majority of developments are very recent. However, clear initiatives can be identified in the areas selected for study: the production control area and accounts department at Apollo and the distribution area and accounts division at Achilles. The computer systems are very recent additions in these departments. It is also a characteristic that new technology is under developed in other departments or its true potential is not realized.

In Apollo considerable staff reductions were possible with the introduction of a manufacturing management system that incorporated a production control system. The remaining jobs were transformed from a manual card system to a computer system that is capable of organizing work and giving update reports as well as controlling stock. At Achilles a comparable system was the stock and distribution control system which also offered considerable savings in jobs but at the same time handled much of the logistics work which made the use of labour in the warehouse more efficient.

The accounts department at Apollo had introduced new systems shortly before the interview visits took place and at Achilles the systems were still in final stages of planning. Apollo had been able to remove much of the work to meet deadlines for a main computer to process the accounts information. New systems allowed the staff to enter data direct to an integrated system which could give updates of accounting information on request. Some regrading had taken place to recognize the additional VDU skills being employed by members of staff. Achilles had plans to use their system to keep closer track of credit and the group of staff dealing with this area were being groomed for upgrading.

Conclusions

The objective of this chapter has been to provide the background information important for a clear understanding of the selected case study companies.

The six companies chosen for analysis represent three areas of new technology development in metalworking, newspaper printing and office environments. These companies further represent three groups of matched pairs in accordance with each area allowing comparison. Furthermore, in each pair of companies comparable areas have been chosen for study of work organization and job definition. For example, machining areas in the metalworking manufacturers, editorial and production areas in the newspaper companies and distribution/stock control areas and accounts departments in the other manufacturing companies studied for their office technologies.

In the following chapters the new technology introductions are considered in each of the companies; the analysis focuses particularly on the work organization aspects of programmes for change referring to the perspectives of workers, managers and trade unions.

Notes

1. There are actually three possible scenarios (a) it is possible to supply engines for the original equipment manufacturers in the vehicle market.(b) engines may be supplied solely for a manufacturers own use in vehicles also manufactured in the same business. (c) a manufacturer may supply engines for use of the business but may produce more and sell them elsewhere to original equipment manufacturers.
2. The supply of agricultural engines is a legacy from a sister company, which is part of the same group as Zeus, and is a manufacturer of tractors.
3. In fact the majority of the CNC machines had been moved here from the main site.
4. Coincidentally the Pearson group owns Westminster Press who in turn own both of the newspapers selected for analysis
5. DTI grants were available as major subsidies to encourage the purchase of Flexible Manufacturing Systems.
6. Notably the company has reverted to staff status for all technician grades and it is possibly for this reason that the term 'technician' is used. However, it is generally a term reserved for production engineering grades.
7. The lesser influence of the unions in the provinces is an important point that was emphasized by both union and management during the interviews.
8. Two major disputes between the NGA and NUJ were at Portsmouth and Wolverhampton. See NGA, *The Wolverhampton Express and Star Dispute - a briefing paper* (Undated) and 'Portsmouth and Sunderland: direct input', *Industrial Relations Review and Report* (1985).
9. Westminster Press is owned by S. Pearson Publishers Limited but operates independently of the group:

   Westminster Press, apart from its 25 newspaper businesses, owns a chain of two hundred retail newsagents' shops and has interests in general printing. It was one of the national chains that caused official concern on account of its propensity to monopoly, as it doubled its share of total weekly newspaper circulation between 1961 and 1974. The parent firm, S. Pearson and Son Ltd., is one of the largest, most diversified and international companies in the world, having interests in banking, engineering, pottery, oil and property. It also owns Chessington Zoo and Madame Tussauds. In 1977 it was the third biggest of the conglomerates owning British publishing houses, with a turnover of 290 million per annum. (Cockburn, 1983 p69)

   Pearson also owns the Financial Times, The Economist and the publishing companies Longman and Penguin (Pearson Annual Report, 1986).

10. However, on the health and safety issues Cockburn (1983 p52) points out her findings that:

    ....this old metal technology had many disadvantages. It was heavy work and tired you out. Often it was hot, dirty and noisy. There were health hazards in working with lead, with solvents and with machinery. The men quite rightly complain of these things. But there was an ambivalence in their expressions, because for many men these very factors were what made the work manly.

11. The changes of managing director and direct entry coordinator proved to be important developments for this case study and are taken up in later chapters.
12. The copy flow structure refers to the channel the news (or advertisements) take through the business. For example the pre technological change copy flow was: reporter - sub editor - news editor - compositor - reader - press. Post technology and direct entry it is: reporter - sub editor - news editor - press.
13. See also Cockburn (1983).
14. MSG1 is the machine shop, MSG2 is the process area, MSG3 is wound products, MSG4 are skilled shops and because they are skilled, although still manual they are considered as indirect employees.
15. Other parts of the business were also considered briefly and this helps to put the overall new technology programme in perspective.

5 The nature and allocation of work

Hypotheses

1 - The diversity of factors governing the organization of work with new technology means that neither deskilled or reskilled work will predominate even in the same firms or industries.

2 - Allocation of work is not always based on criteria of finding the most technically suitable person for the job but political and social criteria may predominate.

Introduction

The first hypothesis in this chapter opens the debate about the nature of the job changes in the case study companies. The second part of the chapter considers the procedures for the allocation of new jobs.

What happens to jobs after new technology is introduced? Some writers approach this question by suggesting a deskilling of work roles (Braverman, 1974; Shaiken, 1985; Noble, 1979), others highlight the new skills that come out of new technology introduction (Hirschhorn, 1984). This is a debate reviewed in chapter two and considered in existing case studies in chapter three. However, despite numerous studies in this area, the effect of new technology remains unclear. Consideration of new case studies is intended to help identify the factors that promote either a deskilling or reskilling of work. In the analysis reskilling is considered in two categories. Horizontal reskilling refers to the extension of skills at an equivalent level of the existing job. Vertical reskilling is the addition of skills at a higher level than workers have been accustomed to.

While there is insufficient evidence to suggest a general tendency toward either a reskilling or deskilling, the most common scenario in the jobs studied was horizontal reskilling. Nevertheless, there were variations in skill levels and the hypothesis is confirmed in this respect.

The second part of this chapter deals with the allocation hypothesis, which considers the way workers are allocated to jobs following new technology introduction. The new jobs may be simply the original job with some additions, or completely new jobs created by new technology. However they are conceived, it is then a problem for managers to decide how to allocate the work to the existing workers or new employees.

Two possible scenarios are, firstly, that job allocation is based on technical qualifications, length of experience and ability to learn new tasks in new technology. On this basis redeployment of existing employees would be expected to be the most common course of action since these workers already possess knowledge and experience of a firms organization and operation. The second possibility considers social and political influences that may shift the emphasis away from the former scenario to the extent that these criteria dominate. Trade unions may successfully apply pressure for an allocation criteria to discriminate against non union members, or members with the shortest service record, should new technology involve job loss. On the other hand managers may attempt to employ new staff as a means of excluding trade union members or removing 'trouble makers.' Other allocation decisions may actually incorporate discrimination against workers on grounds of sex, race, age, or disability which amounts to social exclusion.

Strong evidence exists of the use of political and social criteria alongside technical criteria in the case study companies, to the extent that it predominates in several examples. The allocation hypothesis is confirmed without reservation.

The nature of work following new technology introduction

In this section the effect of new technology on jobs is considered. The decision of job redesign may be as much political as practical. This study is therefore, interested in those factors that may influence a particular outcome in the way that work is organized. These may be based on theories in the job design debate. Some jobs may be subject to sociotechnical design, others deskilled in the form of the labour process interpretation.

Metalworking studies

Artemis, the maker of hydraulic components for aircraft, introduced CNC machinery, speeding up the time it took to machine components and reforming work organization. Workers now take more responsibilities for CNC programming and are flexible across these areas. Zeus, the diesel engine manufacturer, had made similar changes. There is evidence of workers taking on programming tasks and the merging of jobs was being encouraged. However, in Zeus this had less to do with improving service to customers but was more about cost saving.

Vertical reskilling

Artemis engineering reported broader work roles with the advent of CNC and the 'partial FMS.' The manufacturing manager claimed that this was manifest to some extent in part programming of CNC lathes but this was not encouraged and only ever occurred at a limited amending level where workers had picked it up from production engineers, as in Jones's (1982) cases. Lack of encouragement to learn programming contradicts claims that those in 'technician' grade three posts, the highest grade of operator, were being groomed as potential production engineers. The addition of programming skills to operators jobs was also occurring in studies by Jones (1982; 1983) and Burnes (1989). In these studies, and in Artemis, this was to alleviate the shortage of production engineers who normally fulfilled the programming function.

The production engineers did not see this as a threat to their jobs, since they believed that it would take rather more than the training package for the operators to acquire a full repertoire of the appropriate skills. Evidence suggested that they were right. The factory had employed more production engineers with the advent of CNC machining centres, but they had not considered the merging of production engineer tasks with the jobs of grade three operators at that stage. The personnel manager also expressed a strong preference to keep conception and execution firmly separated. So a workforce of operators, resembling production engineers, seems unlikely. It seems probable that if this does happen in the future it will be limited.

The overall effect of new technology in the factory has been to remove some skills and add others. Alongside technology, managers also saw the importance of 'qualitative flexibility' (job flexibility across a range of tasks), and this has also played an important role in redefining work roles. Similarly, work satisfaction has increased for those taking advantage of the module training programme, although some have decided they did not want to take any further training.

The main example of potential vertical reskilling in Artemis was the addition of programming tasks to production workers jobs.

Similar trends are also evident in Zeus where there were other examples, notably evidence of supervisory functions being passed down the hierarchy.

Zeus provided examples of bigger jobs, including potential vertical increases in skills, where responsibility has been pushed down the hierarchy to supervisory levels and to the lower grades. Grade B staff have also been given responsibilities for assisting lower grades with problems they encounter. More tasks have been added to the jobs of the semi-skilled and unskilled in the plant, this will include aspects such as completion of paperwork, use of VDUs, and self inspection. New technology has often played a role in making much of this job expansion possible, for example, there was previously no substitute for the VDU screens.

Programming of CNC machinery was already taking place amongst the top B grade operators but very much on an unofficial basis. As in Artemis what operators learned was based on information gleaned from production engineers. This part programming by operators was not discouraged in the plant and the production operations manager referred to it with some self satisfaction since Zeus was getting something for nothing. What these operatives were remunerated for were the 'manual data inputs' (MDI) required to maintain the functioning of the machine such as changing the tape, checking tool wear, and making adjustments accordingly.

The Era factory of this company was similar in operation but some of the CNC operators here were actually involved in complete part programming and worked with the component from the beginning to the end of the process. This was official, yet still there was no additional remuneration for it. However, these people were permitted to program only for the production of pulleys on two axis machines, which was considered a relatively simple task compared to more complex components on multi-axes machining centres. Elsewhere CNC operators dealt with varying levels of editing and trouble shooting of the production engineers' programs. The operators directly involved in these functions reported that they were happier here than they had been working with conventionals, but felt restricted because they wanted to learn more about programming.

Horizontal reskilling

In the repair and overhaul section of Artemis the use of computer test equipment was employed as an additional testing medium which indirectly reduced the workload since fewer products were returned as faulty and testing was quicker. Staff were freed to do other work, one effect of this was also to make the work more varied. There is also an expansion in the job which, includes the inspection function being been passed to the workforce rather than being handled by specialists. This, the assembly superintendent maintained, made the people doing the work much happier since they would never have known before if stringent testing was the reason that a fault was identified or if it was due to an error on the part of the worker. It was claimed that it is more reassuring to test and inspect your own work and the outcome of this policy has been a fall in the number of rejects and reductions in 'down time.' The possible elements of control and surveillance that could be present in such a system, and referred to by Jones and Rose (1986), were not alluded to.

One of the most important features that restricted flexibility was the old individual bonus scheme. The industrial engineering manager explained that machines were grouped together and technicians are required to move between those machines, in some cases this would involve 'multi-manning.' The problem was a common one in which some workers earned much bigger bonuses than others, so it was difficult to encourage them to be flexible when working on a different machine since this might mean a much reduced bonus. Making the bonus scheme relate to the output of the machine shop as a whole, removed many of the distinctions that existed previously between the pay levels.

The flexibility changes stem from the problems associated with finding enough people with CNC skills, the right company attitude, and a full repertoire of conventional skills. The idea of flexibility was to allow the available skills to be used to their maximum potential, so that operators would be capable of moving to the areas where demand was highest, which could mean operating CNC or conventional machinery.

The other element of the flexibility programme was to make the company a more interesting place to work by making the work more varied, with the motive of capturing more of the skilled workers in the area. Indeed, the industrial engineering manager claimed that workers generally, and particularly those involved in CNC, were now enjoying enhanced work roles. The rigid job classifications that once were maintained (the idea of having a CNC operator, CNC setter and a range of job titles specifying particular conventional machines) have gone, and have been replaced with the general gradings of 'technicians' one, two and three. These gradings reflect considerable flexibility across the conventional and CNC machinery.[1]

Despite the fact that 'labour process' writers argue that tasks have become deskilled with CNC machine tools, it is clear in Artemis that this has not happened, and in fact new skills have replaced the old conventional skills:

....I think there has been a change in the emphasis of skills for example in moving a component from a conventional lathe to an NC lathe the craftsmans skill perhaps making his own tools, the feel that he had in the way that the handle of the lathe moved in relation to the tool cutting - that's gone; but the skill of knowing that relationship in more numerate terms in terms of feeds and speeds has increased (Industrial Engineering Manager, Artemis).

The 'technicians', or operators using CNC machines, generally reported that they were much more satisfied with the new work than working on conventionals. Multi-manning was a particular area of work that they identified as finding more interesting and varied, since it meant greatly increased responsibility. Multi-manning is clearly an area that is directly enabled by new technology. However, one operator suggested that a significant proportion of people would rather work on three or four hour cycles of the same repetitive work, since it carries less responsibility. Similar sentiments were expressed at Zeus. Not all workers were in roles that required them to use NC or CNC, others experienced job expansion in self inspection. This required the use of a personalized rubber stamp to identify that the part of the machining they were responsible for had been completed successfully. There has been no tendency to 'deskill' but in fact the opposite trend appears to have been taking place.

Zeus had been able to reduce the number of grades in their organization in an attempt to reduce manufacturing costs and achieve greater flexibility. The personnel manager explained the effect of this on jobs: 'So rather than have a large number of job classifications, different sorts of job boundaries, we needed to make jobs bigger so that we could make the workforce more flexible and reduce the total numbers.'

In some jobs this means retraining for new skills, and in the maintenance area, for example, this is a direct result of new technology, since the degree to which a job was mechanical or electrical was becoming increasingly blurred, a development also recognized by Rainbird (1988) and Jones (1988). So it was decided to merge the two jobs by interchanging the skills required, although the skilled workers still remain specialists in their own fields. This was of considerable advantage to the employees as well as to Zeus. The staff not only take more interest in the work and have more responsibility but also gain a qualification valuable outside the company (see allocation hypothesis this chapter and the demarcation hypothesis in chapter seven).

Deskilling

One assembly worker at Artemis made his highly divided and closely timed job sound like scientific management organization. But the difference was that the small repetitive parts of the job were performed by one worker as a range of tasks, and not divided up between workers as individual tasks. However, the fragmentation clearly bothered operators who would have preferred more control over their own work. This shattered the management's image of a harmonious workforce. There was also concern about the quality of the products since timing encourages short cuts which may lead to problems.

However, the distaste of timing was mellowed by the tendency of the operators to rotate allowing them '....to get the feel of the whole component....' Although there was clearly a distrust of flexibility as a possible covert means of rate cutting. Nevertheless, the management aim was to continue training with a view to further increasing 'qualitative' or 'functional flexibility' (Atkinson, 1985; Atkinson and Meager, 1986). Nobody mentioned that these flexibility measures may also embody elements of control but for Palloix (1976) and Pignon and Querzola (1976) job rotation and enlargement are consistent with this possibility.

Like some workers at Artemis, at Zeus one of the senior shop stewards, an unskilled assembly worker, argued that he was happier with a simple continuous series of highly divided tasks, rather than the present scenario of being moved from one place to another and performing quite different jobs. As at Artemis, the trade union had cooperated with changes for the good of the business although one senior shop steward saw new technology as causing job loss, another as bringing repetition and boredom consistent with labour process writers (Braverman, 1974; Shaiken; 1985; Noble, 1979). CNC machining, for example, was described as button pushing, suggesting an ignorance of what is really involved. Work by Wilson and Buchanan (1988) and Batstone *et. al.*,(1987) has confirmed that this is a common belief amongst conventional operators and an issue of annoyance for the CNC operators.

The appearance that work is easier overlooks the additional responsibility or new knowledge that may be required. For example, computerizing of certain pieces of equipment now means that operators are able to perform functions that previously would have required supervisory intervention, but because they now have the information they require it is easily accessible. Hence control has become more machine oriented, although the supervisory function is still recognized as being important in the organization. Jones and Rose (1986) also found less need for direct supervision due to the surveillance capability of new technology. This denies a labour process model, although control over work may be attained by an alternative means to deskilling (Pignon and Querzola, 1976).

Summary - metalworking studies

Artemis has experienced a series of work changes. In the case of CNC, these included multi-manning and unmanned running. These are both areas enabled by CNC equipment, and workers were particularly satisfied with the work variety of multi-manning. Elsewhere quality inspection was an addition to all skilled jobs. There is no evidence that deskilling has taken place but the use of the stopwatch, and fragmenting of tasks as components within jobs, rather than seeing jobs as a whole, caused dissatisfaction for some workers.

Zeus provides a very good example of a company that has successfully achieved the expansion of jobs and addition of tasks without having to pay extra for them. This point is taken up in the next chapter. Zeus's changes in jobs are generally to add new tasks to existing jobs. A good example of this is the transfer of electrical skills to mechanical craftsmen, and mechanical skills to electrical craftsmen. But at semi-skilled levels CNC operation has meant the learning of programming skills, quality is now the responsibility of individuals not a central department, and less supervision is required. Generally workers reported greater satisfaction.

The metalworking companies display clear attempts to expand key jobs particularly with CNC machinery, but they stop short of ceding full responsibility to the workers. CNC operatives in both companies expressed a preference to CNC work over that of conventional operation. However, in other areas, workers recognized the benefits of highly detailed work, arguing that this meant there was little responsibility, and as a result less job related stress. There was strong evidence in a range of jobs of trying to achieve flexible operation, which would then relax job definitions, blurring old demarcation boundaries and solving difficult allocation problems. For example, in the maintenance function at Zeus. The same policy often allowed Zeus not only to enlarge jobs, but also to avoid paying for the addition. Workers in both companies generally suggested that they were happier with more varied, responsible work, but others suggested that repetitive work was better because it offered less responsibility and stress. The evidence here is against any one outcome, but consistent with the hypothesis reflecting a variety of skill levels.

At the newspapers the move to greater flexibility is evident, but only after the new technology change to a full direct entry system. Here new skills are taken on by editorial staff but production jobs are deskilled and often reallocated.

Newspaper studies

There is little difference between the two newspaper companies in the overall effect of the changes, although there are differences in the way the outcome was reached. In both companies deskilling on the production side of the business is met with reskilling on the editorial side. These changes are driven directly by the technology that was employed.

Vertical reskilling

Opportunities for gaining higher level skills were available for production workers in both newspapers. This could have occurred with the redeployment of production employees in editorial jobs as was the case in a study by Smith (1988). This would have meant a certain degree of retraining and building on existing knowledge. However, this potential reskilling was never realized because managers were reluctant to allow this form of redeployment on political grounds, an issue discussed in the next section on work allocation.

Horizontal reskilling

Mercury Newspapers recognized that it was important to retain 'readers.' These were NGA composing room staff, with the responsibility of reading the pages of copy[2] to ensure that it was accurate in spelling and grammar before going to the press. The retention of readers is specifically to reduce the loss of revenue in advertisements, non payment of the newspaper due to errors in advertising was apparently becoming a costly problem. Reading is redundant in Hermes, management claimed that the possibility of errors has been reduced by avoiding the retyping of copy, although consideration was being given to the retention of one reader since it was accepted that retyped, contributed copy, would quite possibly contain errors.[3]

The changes for editorial staff have been most marked for the sub editor who now has more responsibility as the final line of checking in the newspaper. The sub editing function has become more important to ensure the copy is submitted correctly. Interestingly, the sub editing of copy on screen is reported as being significantly slower than sub editing on paper, and the newspapers had begun to take on more staff in this role.

The small change to reporters' jobs attracted a pay increase, but also was claimed to have made the job more efficient, allowing the rapid editing of text on computers before submission to the sub editors queue.[4] The deputy editor at Hermes was at pains to highlight the increased responsibility that the reporters were now taking on since they could no longer rely on a succession of 'checkers' to rectify their mistakes. He was also determined to divide the work so that menial work was left to production workers. This included inputting functions, in which he felt journalists would be wasting valuable time and skills. The trend is indicative of a general change evident elsewhere to deskill production workers and reskill editorial staff and segregate these groups.

Deskilling

The production manager at Mercury identified the main objective of new technology as removing the double setting of copy, and in this way reducing composing room costs. Which introduces the need, if not the motive, to deskill existing NGA employees.

In the change to a production take photocomposition system at Mercury Newspapers, the possibility for greater flexibility was apparent. This was not realized due to long established union demarcations. What was achieved here was the significant deskilling of key NGA staff who previously performed jobs that drew on a long apprenticeship. The exclusivity of their skills gave them control over the newspaper, no one could step into their jobs and take their places unless they had the skilled knowledge required, and were members of the NGA. The union had constructed elaborate barriers to entry, allowing only those who were members to get jobs (Cockburn, 1983).

Production take photocomposition sufficiently deskilled the compositors' work for them to be replaced by other less skilled workers although at this stage the structure was maintained but must be seen as part of a step by step approach by management to slowly gain control over the industry. In an overwhelming number of cases similar trends are apparent (Cockburn, 1983; Martin, 1981; Smith, 1988; see chapter three). The real threat to the unions did not occur until the company made the decision to use direct entry technology, which largely ended the need for separate composition of text. Since the change to direct entry photocomposition, Mercury has eliminated union control over demarcations, and begun to use production assistants to input copy. These people work four hour shifts within six month contracts. They are exclusively women with typing skills, and are effectively on call if the newspaper falls short of staff. In this way a once highly skilled job is turned into a detailed routine job.

There is little difference in Hermes, except this group of employees have the more down to earth title of copy typists and are employed in both editorial and in the 'tele ads' area, dealing with advertising copy. The two sets of staff have to be trained, the former in using a system that allows the input of copy in an advertising format, and the latter to operate a database system.

The progressive change from highly skilled NGA workers, to unskilled or semi-skilled workers, contrasts with the role of the journalists whose jobs have actually expanded to include the direct inputting of their own copy. There is a clear trend toward the deskilling of NGA jobs, and at the same time extending worker flexibility (in the quantitative sense) almost in a celebration of the demise of union control. It also exploits the cheap female labour that is casual and numerically flexible (Walby, 1989; also see Cockburn, 1983; Phillips, 1983).

NGA staff themselves face deskilling or the loss of their jobs. The other option for a few is to remain in their existing jobs, setting the more complicated work such as display advertising. The work generally has been deskilled but the NGA FOC expressed his concern that the extent to which the newspaper could operate with new staff was perhaps being over estimated:
'....what they are doing now is deskilling, bringing in these female inputers, non union. But once you get into technical adverts, on screen, there is still a lot of skill there....' (NGA FOC, Hermes).

One of the overseers in Hermes expressed similar concern that the idea of reducing costs was quite false because there were likely to be problems with new staff and therefore costs probably would not fall. For example, in Rainbird's (1988) study, old staff who had lost their jobs had to be recalled to train the new staff in tacit functions known only to the old staff. In both newspapers the retention of compositors in areas such as display advertisement 'make up' suggests managers may have been at least partially aware of this problem.

Summary - newspaper studies

For both companies the inevitable control motive embodied in the decision to go forward with further technology changes, has meant radical changes to production jobs such that the old skills are irrelevant for the new processes. The deskilling of the work has been accompanied by a replacement of the traditional staff with people with fewer skills, who are paid considerably less. There was a feeling that the editorial department was more in control of producing the newspaper, and the staff in these departments had new tasks added to their jobs and were paid more in recognition of this.

Opportunities for vertical reskilling were not taken up in either newspaper. This would have meant introducing production workers into the editorial areas which was clearly politically distasteful for managers and editorial staff (see next section).

The use of editorial staff, with enlarged jobs on the one hand, and deskilled production work carried out by workers on casual and short term contracts, on the other, suggests a trend similar to that identified by Atkinson (1985) and Atkinson and Meager (1986). That is, the existence of a workforce divided between functionally flexible employees and numerically flexible employees (see chapter two). However, the experiences of the newspaper companies also demonstrate a tendency toward increasing control over the core group of journalists, holding them more accountable for errors. It was very common in the past for editorial management to lay the blame for errors in the text on the compositors. Journalists had lost their scapegoats, and were now directly accountable for their mistakes and easily identified. Inconsistent with the changes in metalworking, there is more evidence of deskilling, but consistent with the hypothesis, even here, reskilling has also taken place.

Like the newspapers, the manufacturing companies studied for their clerical jobs, demonstrate both work organization for increasing control and for achieving flexibility. The tendency to keep the skilled staff away from the mundane work was present in both companies.

Office studies

Apollo reduced the amount of mundane work that staff were undertaking, and in general, skill levels have increased. However, at the same time the introduction of new technology has seen the creation of two layers of jobs, some that are being reskilled but others that are either deskilled or remain at their previous low level of skill. For example, instead of manually filling in forms the same function is now carried out on computer. Achilles used new technology to avoid a burgeoning of clerical staff, and to facilitate more efficient information transfer. More tasks were added to jobs, and in some cases, supervisory control receded and staff dealt with new responsibilities. However, this was often met by an increase in control enabled by the computer systems.

Vertical reskilling

Although Apollo display no evidence of vertical reskilling, the possibilities for this occurring were similar to those at Achilles. Managers had decided not to cede supervisory authority to workers, while at Achilles there was some evidence of this tendency.

On the accounts side of Achilles, the computer system that was required had not yet been implemented, but there had been some clear decisions made about how the work was to be organized. The management in this department had decided to make a clear split between the staff, some jobs were to be menial and would involve the routine reconciling of invoice details. The division of these functions would then free key credit control staff to do the 'more responsible work.' However, other jobs affected are those of invoice clerks who currently produce invoices on typewriters but have to refer them to a supervisor. The technology changes will make their jobs easier because the information they require will be more accessible, while at the same time they will need to refer less to a supervisor and be able to make more decisions by themselves.

Therefore an important part of the change is the tendency to push responsibility down, relying less on the supervisors and more on the clerical staff themselves. Another group of clerical workers are firmly marginalized and have to deal with the mundane and routine work. The most important point is perhaps that there is no departmental trend towards deskilling, but the nature of change is very job specific. Hence, the credit manager at Achilles explained, the job of the accounts payable clerk would involve more skill because the employee would be required to decide, on the basis of information given, the amount of time a customer should be allowed to pay. On the other hand, the post of accounts receivable clerk was demanding less responsibility than before since it involved entering data into the computer, that then handles the calculations that would previously have been manually executed. Studies by Storey (1987) and Batstone *et. al.*, (1987) identify similar trends.

Horizontal reskilling

The personnel manager at Apollo described the overall result of new technology as reducing staff numbers, and as far as job design is concerned, reducing the mundane work that used to be so common. There were examples in his own department where he described the collation of clock card data as a manual process based on a card system and requiring some six people to operate it. Now that the computer collects all the information only two people are required, they are able to access and withdraw the required information from the 2000 clock cards quickly and easily. Elsewhere the typing and retyping of reports to win business are now done on word processing equipment. Reports often require more than one set of changes, and word processors take away the need to retype the whole report. Reports can be called up and edited on screen. The function no longer requires a typist but can be handled by the technical staff themselves. It also means that relevant text can be recalled and reused in later reports. These changes are common, and are consistent with studies by Wainwright and Francis (1984).

The work in the accounts department, prior to the use of computers, was based on manual writing, followed by a separate punching of this data into the mainframe computer in another department. The system was very labour intensive, and the introduction of a localized computer system, allowing the direct entry of data, reduced the numbers employed. At the same time the workload has increased significantly.

However, it was in payroll that offered one of the most interesting examples of the way that the organization of jobs has changed. Previously, a large amount of the payroll work was handled outside the firm, and the proportion that was handled inside had to be sent to the London head office for processing. Computerization has meant not only bringing the monthly payroll back into the firm, but also handling other aspects of work such as SAYE. The staff all work on different payrolls, but have become more flexible and are rotated to different jobs in the department. They have much greater volumes of work to clear, and the emphasis is on spreading out the routine work that remains to all the people in the department, rather than concentrating it with one or two individuals:

> Their jobs have expanded a lot now....they all work as a team whereas when it was done manually you would probably be stuck with the same part of the payroll all the time because it took so long. Variety would be very small but now we have the variety all the time (Payroll Administrator, Apollo).

At Apollo overall, the emphasis has tended to be on improving the quality of service that clerical departments could provide but in many cases there was undoubtedly a benefit in the improvement of the quality of the jobs.

At Achilles the systems manager claimed that there was an identifiable increase in the flexibility of staff to meet the peaks and troughs of the company's business. The similarity of the VDU systems means that often staff are able to learn jobs quite easily. The personnel manager also pointed out that there was a great advantage in the newly developed flexibility of the staff that was afforded by new technology. The use of computers had meant a reduction in personnel, but alongside this, had allegedly created more interesting jobs for remaining staff, and provided the equipment with which a far better service could be offered to other divisions than had previously been the case.

The distribution and customer manager outlined a technological transformation, that had not only meant greater flexibility but one in which the very structure of the workforce had changed: '....the general trend has been a vast reduction in the number of lower grade jobs, there are still some lower grade jobs but new jobs are being created and there's been a general drift towards less people, but doing bigger jobs.'

This suggests a shift to an increase in skill levels as lower grade jobs are forced out. Clearly, the trend towards larger jobs, and the flexibility of the clerical staff are interdependent features. The facilitator is the new technology which has also introduced the trend of altering the nature of the manual work. The staff at this end of the spectrum remain involved in highly divided tasks and are either unskilled or semi-skilled.

Deskilling

In the invoice payment area of Apollo the supervisor described some of the new jobs as being more varied and enhanced. However, of some, the contrary was true. Some staff enjoyed no apparent enhancement, and they had to deal with the mundane task of inputting data into the system. Unlike the payroll section, the inputting of data was handled by staff members who did nothing else and would never see the beginning or the end of a process. This cannot really be seen as a deskilling since the previous role was just as mundane, but involving form filling rather than computer input. Other staff and managers in the area reported that the work had become easier.

Importantly, the work was also reported to be much less interesting and requiring virtually no human interaction, whereas before information often had to be sought by visits to different departments. Now the information could be found recorded either on the department's computer system, or by accessing systems linked into the accounts computers. Wainwright and Francis (1986) found similar developments, and control systems can be identified like those referred to by Pignon and Querzola (1976).

It is difficult to argue deskilling has occurred here without an intimate knowledge of the jobs themselves, but it is clear that the computer system was able to maintain closer control of the workforce in two ways. The first by physically keeping the employees in one place where they could be supervised. The second because the computer system was capable of storing information about who was responsible for work, and when it was done, so that individuals could be identified if errors were made. Hence an apparent enhancement of work may embody covert control systems (Palloix, 1976).[5] The contrast between the payroll section and the invoice clearance area is interesting and may be explained by the greater autonomy in payroll. The financial accountant ruled out any other form of organization outside the design style in invoice clearance, such as the passing down of responsibilities to lower graded staff, because this had been practiced and had failed at his previous place of work.

In the manufacturing services division of the organization, the job of the progress chaser and order processing staff had been greatly simplified by new technology in the form of a production control system. The shift from manual records to computerized ones has meant that the previous problem of lost details, or details that were at a separate stage of being processed, would be eliminated because the accessibility of the information was so much improved. However, despite achieving a pay increase, the work had become simpler because the system was easier to access, but it was not deskilled since similar skills were required. The argument that the tools had changed and not the job, was echoed here. The reduction in staff had meant an increased workload for those remaining in the functions.

These jobs require a heavy use of VDUs, and the use of VDUs also cuts across the boundaries between direct and indirect workers (see Coriat, 1988; Arthurs, 1985). All the indirect staff use the system and work with paperwork that is generated by the system. Some people who are 'directs' also use the system but as a rule this does not include operators. Passwords restrict access to those who are entitled to see information, and at the higher levels of access the changing of data is also permitted.

In many cases the manual jobs in the distribution area of Achilles have become deskilled, since the computer system instructs workers where to go and what to pick. Things they themselves would have had to decide, without the aid of the new technology, prior to the stock control system's implementation. There are some elements of interest introduced in the slightly wider repertoire of tasks that a manual worker can undertake related to the use of the system.

In common with the metalworking studies, the clerical trade union (APEX) saw new technology resulting in deskilling and job loss. There was a general, simplistic view that technology can perform functions automatically thus making work easier and reducing the need to employ as many staff. It is clear from the studies in Achilles and other firms that the relationship is rarely so simple.

Summary - office studies

Apollo demonstrates the way that the deskilling and reskilling of work can exist together in the same company, and even in the same department. Staff generally reported greater variety at work, but some work had become more mundane and simplified. Achilles show more evidence of increased worker flexibility resulting in less need for supervision.

At Achilles, like Apollo, more cases of simultaneous reskilling and deskilling are found. But here there are also significant changes in the use of systems as control mechanisms, meaning that traditional forms of supervision has become less important.

When jobs are repackaged following the introduction of new technology some aspects are new and involve more skill, other features are subject to deskilling. For example, in accounts departments more skill is required because, given better information availability, staff are required to make decisions. On the other hand, functions previously requiring manual calculation, are dealt with by the computer. Features also occurring in studies by Storey (1987). In these cases the hypothesis is only partially accepted. There is evidence of reskilling, but strong evidence of control motives also exists.

In the clerical cases, the tendency to introduce new technology systems to deal with larger quantities of work, and to create more possibilities for information accessing and cross referencing, are particularly evident. There is an increasing trend, with new technology, for direct labour to take on indirect clerical tasks in the course of work, manifest particularly in the need to access computer databases.

As in the other cases the new technology can mean both the addition or loss of skills, and this is consistent with the hypothesis. However, in both cases it often means more control, since the kind of work now performed is such that even if it does require more skill and responsibility, computer systems can do away with the need for interaction with other people because the information is readily available on the computer. This cuts down on the physical movement in the department, and on the site, and may also mean that managers can identify individuals errors by using the system as a means of surveillance (see Jones and Rose, 1986).

Decisions about the way jobs are repackaged, following new technology, form the basis of studies of job redesign and work organization. What comes next are decisions of who to employ in the new or altered posts. The next section pursues these issues.

The allocation of workers

The introduction of new technology systems inevitably means changes to jobs. In some instances there are only minor task additions or subtractions, there may be no major organizational change. However, other cases exist where jobs change radically and may involve considerable reorganization. Some tasks become redundant, others are created. This change process requires decisions about the deployment of people, and it is these decisions that this hypothesis is based on. Like the design of jobs there is considerable room for choice. Political and social factors influence the choice of people for jobs, in similar ways that social and political decisions affect the choice and deployment of technologies and the designation of jobs.

Metalworking studies

In the metalworking cases, studies focused on allocation criteria for jobs in machining and assembly areas. Artemis and Zeus were both concerned that older workers would be a problem to train and would have difficulty assimilating new technology. This policy at Zeus conflicted with a trade union view of promotion by seniority. Zeus was also looking into the possibility of using more contract labour rather than employing people permanently.

Social and political criteria

At first, Artemis revealed little in the way of any allocation criteria. Workers, the production director maintained, were trained in CNC and were simply transferred to the new machines. Others, untrained, were offered the opportunity to get involved in the module training programme. The machine shop superintendent maintained a process of 'personal negotiation' during this period to allay fears of the adverse effects of new technology. Although the company did take the opportunity of removing some of the older staff, 'trouble makers and dead wood.'[6] Nevertheless, the production director claimed there was no loss of staff attributed to unsuitability or incompatibility of employees with the new technology, but the attitude towards the older employee seemed to contradict this.

For both inclusion in the training programme and to serve as CNC operators, younger people were preferred. The manufacturing manager felt the older employees saw little value in training and claimed that '....you can't teach an old dog new tricks....' The worker's age and 'suitability' was a recurring theme in interviews with managers. The industrial engineering manager described a situation where older workers had left with the removal of old machines, unwilling or unable to retrain for modern machinery. Allocation criteria, described by the industrial engineering manager however, did not seem to necessarily exclude older workers:

> We found that we needed two or three elements, we wanted people to have basic overall engineering skills, we then wanted them to understand about us as a company and about our products and then thirdly to operate CNC machines because we were heading towards CNC machines, so the workforce should be able to use CNC machines (Industrial Engineering Manager, Artemis).

Nevertheless, at Artemis, the management had stereotyped older workers as incapable of further training, or working with new technology based on their experiences.

At Zeus the attempts to increase the flexibility of the workforce had revealed severe weaknesses in the ability of some employees, and made the management realize that it was not possible to assume that every worker would be as flexible as the next: 'There are certain people in any environment that are only capable of doing certain jobs, untrainable, very difficult to train them in different jobs (Production Operations Manager, Zeus).

The company dealt with this by simply accommodating those workers who were not capable of retraining. They received the same benefits as the workers with enlarged jobs, but it was accepted that they would maintain the narrow job they had always done. For example, in one machining area five setters and twenty five operators have become fifteen setter/operators, with some operators remaining who it was felt were not capable of making the leap to perform the range of tasks required for setter/operator. The least flexible group were the most vulnerable and obviously earmarked for removal if there was another round of redundancies.

No new staff were used by the firm, the last time new workers were employed specifically to operate the new technology was eight years previously. The only new staff now are apprentices and graduates. The company had thought through the possible future need for staff and the production operations manager claimed that in the future: 'We would....take temporary labour on, short term contract labour rather than full time labour.'

This is likely to have only been possible in lesser skilled jobs. It became clear that there was already a considerable degree of contracting out of some operations such as the cutting/grinding function which was previously a central service. Although unconfirmed, the company seemed to be thinking in terms of simplifying tasks with the use of new technology, and then planning to rely on casual labour in the new unskilled jobs. Not surprisingly the trade union was opposed to the use of any temporary or contract labour.

The management had been quite systematic about the way they handled their employees and there were few people in the organization who had not had expanded job descriptions. In some cases the job description was expanded prior to tasks being added to the jobs, in anticipation of future changes, and particularly to offset union claims for pay increases (see chapter six).

Higher gradings were bestowed according to the job description, and whether or not the tasks specified in that description were being fulfilled. Where workers had to be selected this was done on the basis of supervisory knowledge in the sections. When CNC operators were required, existing grade B staff were asked to volunteer and then the 'best of the volunteers were singled out.' The production operations manager at Era expressed concern that there might be too many 'fifty eight year olds' volunteering for CNC posts but, fortunately, the problem never materialized and those eventually selected were 'the people most capable and willing with the greatest potential.' The statement is characteristically value judgemental and vague. Similarly the criteria the section supervisor put forward were equally poorly defined: 'We were looking for people with a good average intelligence level, who were good time keepers and were basically wanting to work....decent workers, company minded, who didn't lose a lot of time and wanted to get on....'

In a similar way people were called to volunteer for extra training. Those who volunteered tended to be young, although some older workers were trained successfully, perhaps bringing into question the link that managers had drawn between age, flexibility and ability. However, the trade union were concerned about the possible embarrassment of workers, and insisted on a clause that stated nobody should be embarrassed by not being able to handle the training. Strong feeling in management circles maintained that this clause was protecting some workers who were quite capable of handling the training but were too lazy to try.

Redundancies were not handled in this way because, the production operations manager explained, the people who volunteer are not always the ones the company wants to get rid of. Sets of questions were prepared and the workers were marked against company orientation as well as ability, a formula that the personnel officer described as 'suitability criteria.' Those with the lowest marks were the ones to lose their jobs. Politics clearly plays an important role here. The 'company orientation' of a person has little to do with practical ability and the best qualified person for the job, but embodies a view informed by political factors outside technical qualifications.

The trade union claimed some involvement on manning levels, but were unable to persuade the management that the most satisfactory method of allocating higher positions was length of service, and not on the management favoured basis of merit. The company had established a new allocation criteria for new technology which implicitly questions the suitability of older workers and is biased towards the younger better qualified, but inevitably less experienced candidates.

The trade union's insistence on a 'seniority rule' was clearly not very realistic, and alternative suggestions, to tackle the perhaps rather subjective and value laden way in which the company were allocating the jobs, were absent. Both a lack of awareness and a degree of vested interest was involved. All the union officials interviewed were over fifty five years old. The senior steward at Era reported the same management attitude towards the older worker claiming that nobody over the age of fifty was wanted for training. He put forward similar arguments as those of his union colleagues elsewhere in the company, that the senior worker should have preferential treatment.

Technical influences

Those involved in deciding to whom the jobs should be allocated at Artemis were mainly the personnel manager and the functional manager, which often meant the manufacturing manager. The choices they made seemed to be based on exacting skill requirements of a City and Guilds apprenticeship, although one grade three felt that the job could be done by less well qualified workers. Both the insistence on high level qualifications, and age biases, are policies that contradict the regular references to the skills shortage and a difficulty in finding qualified staff.

After appointment the progression through the grading structure is achieved as different training modules are completed. The manufacturing manager tried to give the impression that Artemis had acquired a sense of social responsibility, recognizing that all companies should make a contribution towards training. However, the truth was clearly that the skills were not available on the open market and it was therefore essential to undertake internal training.

Moreover, a change in the grading system necessitated each worker being allocated to a new grading level, conducive with the new system. In the machine shop the allocations were based largely on the opinion of the machine shop superintendent, and a degree of inconsistency brought confusion and dissatisfaction amongst some workers. While as a general rule the most highly skilled, grade three operators, worked on CNC machines there were a number of exceptions of older employees who were performing conventional functions and had never worked elsewhere. The inflexibility of this group perhaps served to confirm the companies view of older staff. Nevertheless, the gradings bestowed here proved to be in recognition of the length of service and seniority as well as the skills they brought to the workplace, although conventional. These cases were not common, and for the most part younger workers, skilled in CNC, and 'more flexible and adaptable', occupied the higher gradings.

In the new build assembly area the problem was more acute and not associated with new technology. Having removed its semi-skilled workers dealing with specialized functions, replacement with skilled workers with broader knowledge meant that it was necessary to go outside and recruit new people. Redeployment of existing staff was seen to be unrealistic. The existing workers were mostly lost in early retirement and voluntary redundancy.

Artemis had particularly high technical standards for new employees which was possibly serving to exacerbate the problem they were experiencing in skills shortages. Employing highly skilled workers had not precluded the need for further training, and the choice of workers for training were based on social criteria of age as well as technical ability.

At Zeus, as in all aspects of training, the skilled maintenance area had established a no embarrassment clause via the union to protect any workers 'incapable' of benefiting from training. The maintenance supervisor was fully behind such a clause, and where the older element were concerned the decision to train for multi-skilling was left to individual discretion. Importantly, there seemed to be no attempt by managers to dissuade older workers from training. Perhaps this was because the prospect of having a fully trained multi-skilled craftsman was attractive to other companies, who would pay well to secure staff of this calibre. The shop steward at Era had suggested that some 'head hunting' had already taken place. This is a problem recognized by Jones (1988) in a discussion on the multi-skilling of craftsmen. The rationale behind training without age restrictions may then be based on the expectation of the loss of some staff, therefore it was important to maintain trained older staff in case they would be needed.

Summary - metalworking studies

Both studies provide evidence in agreement with the hypothesis. Artemis, allocating jobs on the basis of high technical qualifications and age, and then training internally, seemed to contradict a fundamental problem which had become a grave concern in the company. The shortage of skilled workers had not made the company rethink its policies towards older staff, or lesser qualified staff, who would be trained 'in house' in any case. At Zeus the nature of the workforce was different but the managers harboured similar biases.

Zeus allocated workers to jobs on the basis of their age, which in turn were based on ideas that older people were difficult to train. Managers and trade unions differed on the way workers should be allocated. The trade union view was that the most senior should have the best jobs, but at the same time insisting no one should be embarrassed because they could not handle a particular job. While managers were looking for the most suitably qualified people they had included criteria that was overtly political.

At both of the metal manufacturers there was strong bias toward the use of younger employees in the use of new technology at the companies. It was common for the companies to discourage older workers from becoming involved in training programmes, and there was an avoidance of using them in new technology, based on a belief that they were inflexible. However, there were areas where older staff had been successfully trained and this seemed to question the managements hypothesis. Undoubtedly another reason for not using older workers was simply because once trained they would be of use to the company for a shorter time than younger people. Hence, young people offer a much more favourable return to investment in training. This however, is only the case where the younger person chooses to remain with the firm. It is probable that a younger person would be more likely to leave in pursuit of better paid work, and in this respect, it may be that an older employee is a more secure investment.

Zeus's use of new technology was possibly the more controversial of the two metalworking companies, they were becoming very keen on the idea of using contract labour. New technology in some areas offered an opportunity to simplify jobs so that casual workers could be used to deal with these tasks. At the other end of the skill spectrum workers were being multi-skilled so that they could handle a range of functions. This means that allocation need not be so precise. If a worker is capable of a range of tasks, then that person can be placed in an area in which he/she can be called upon to handle a variety of tasks that go far beyond the original confines of the job descriptions. Wilkinson (1983) identified similar trends. Intentions at Zeus to use contracted labour also reflect the Atkinson (1985) model of numerically and functionally flexible groups of workers divided to perform different roles.

In the newspapers there are clear divisions between two groups of workers in editorial and production departments. Policies of allocation have become politically charged and seem particularly to be biased against production workers. There is also evidence of gender typing.

Newspaper studies

In the newspapers the production staff are subject to both social and political criteria. The rejection of NGA workers for work in the editorial department seems to reflect entrenched and embittered attitudes.

Social and political criteria

The editorial management at Hermes decided that anybody wishing to move into journalism in the department must successfully complete a test, as well as undergoing an interview. There were six compositors who were interested in transferring to editorial at Hermes, but only one seemed to have made it through the vetting procedure. Later, even this person proved to be below the exacting standards of the department and was rejected.

The NGA father of chapel was unhappy about the difficulty that had been experienced in getting even this far into the editorial department. No other department applied the same kind of stringent testing and NGA staff interpreted this as meaning there was no intention ever to consider NGA staff for journalists' work. Passing them through a selection procedure was merely tokenism to appease the unions:

> ....what we said to the NGA was we would consider them for other vacancies which arose in the company and some expressed an interest about coming into the editorial department....[they failed]....to meet the editors normal criteria for employing journalists....The cynical would view the whole exercise as somewhat fraudulent and the charitable would say out of a small organization such as....[Hermes]....it was perhaps asking too much to find that many people who....had the skills that could be changed (Deputy Editor, Hermes).

The same deputy editor doubted that compositors would meet the criteria laid down by the advertising department either. He explained that the newspaper was not prepared to create jobs specially for compositors or take them into existing posts as 'coasters.' But it seemed to be the case that the company did not want ex-compositors in the business or at least outside the production areas and inputting jobs in advertising and editorial. For the NGA FOC, there were clearly undercurrents of bias in operation in the editorial department against compositors, and these extended over and above questions of ability. There had always been traditionally sour relationships between journalists and compositors, based on resentment of the power compositors had over the newspaper, and the disparity in earnings (Smith, 1988).

The NUJ FOC pointed out that editorial staff in the department saw the prospect of compositors training for journalism as causing denial of entry to junior trainees. But they would also have had to cope with a trainee who was paid more than many of the experienced and highly skilled editorial staff. The editor would certainly have stirred up some internal industrial relations problems amongst his own staff had he decided to employ compositors. After the editorial rejections, NGA compositors were less enthusiastic about applying for jobs in the editorial and advertising areas for fear of having to suffer the embarrassment of rejection.

Compositors were unhappy about the redeployment issue, but were also making reference to cases in which newspapers were re-employing compositors, after they had realized that using unskilled staff was not so simple. Similar scenarios are reported by Rainbird (1988).[7] The NUJ Father of chapel suggested such claims were exaggerated, and in any case the company was quite prepared to put up with the inconvenience of six months of reduced standards, to ensure that the NGA's control over the business was broken. However, the fact was that the newspaper had offered large cash benefits as a bonus to persuade some of those volunteering for early severance, to remain in the company and cooperate with the changes for a temporary period. This at least adds some credence to the compositors' argument, but unfortunately does not support any long term need for traditional skills.

Mercury's editorial department seemed to be adopting the same stance as the one at Hermes. Reports from the NGA at Mercury (as well as commentators at Hermes), suggested that the managing director had insisted that there should be editorial opportunities available for compositors, forcing the editor's hand. According to the NUJ father of chapel, this provoked the editor into demonstrating his authority, thus becoming determined to adhere to his policy of not employing compositors, and using reasons similar to those at Hermes. The NGA father of chapel, who was employed as a reader, claimed that the exclusion was political and not based on ability:

....they are putting all sorts of blocks in the way of people. I think the option to join the editorial staff, whilst it's there in theory, has been made virtually impossible (NGA FOC - Reader by profession, Mercury).

Apart from tests in grammar and spelling, another barrier to the compositors entering journalism was the prospect of training away from home for long periods of time. The NGA father of chapel saw this as a realistic proposition for young people, but not for middle aged men with families. It was argued that the editor only had to produce some poor test results, or report that a person had interviewed badly, to cut short any application. However, consistent with his editorial colleagues the NUJ father of chapel presented a derogatory view of NGA employees and was clearly opposed to their redeployment into the editorial department:

You see what has happened on other 'papers, people have come across from the NGA to be production journalists....to be sub editors, to work on page layout and all the rest of it....we were in agreement in Mercury that, that has really created a second class journalist, someone who didn't have all the skills, they would just learn to do that particular job (NUJ FOC, Mercury).

However, despite this claim, people with existing knowledge of the newspaper and some of the skills, obviously have definite advantages over completely new staff. There was more general union concern about the redeployment issue because of the way people were selected. It was a positive feature that there were more opportunities, and managers were often happy to put somebody on trial. But the way the selection took place was management contrived and meant that people were being channelled into specific areas and being advised to apply for particular posts, not really making their own choices. Furthermore, there was no opportunity for NGA involvement or regulation in this process.

Elaborate barriers to entry had been established at Hermes to prevent production workers being redeployed in editorial areas. In Mercury the same problems were observed. Interviews also brought out the issue of gender typing in newspapers. For example, the production manager at Mercury was adamant that it was not simply the jobs but also the nature of the work that had changed:

> And when we change to cold composition it's a feminine job it's totally different, it's clean, it's quiet, it's not the same kind of thing, it's a different kind of keyboard. The other keyboard was a big keyboard with big keys and different layout. It's a feminine QWERTY keyboard....there is an important transformation from metal weights to cut and paste and eventually even that will go....I don't say big strong women couldn't have done it but....generally women would not have wanted to do it. Nowadays it's a very light aluminium plate and there's no reason why she couldn't do it and there are more women who have come into this industry and men will go out of it. The industry has changed (Production Manager, Mercury Newspapers).[8]

Cockburn (1983) found that the change to production take photocomposition failed to displace men from their work roles due to male dominance through NGA control. Women were excluded in NGA areas of Mercury and Hermes until the pre-entry closed shop ended. Direct entry has meant a radical change of the workforce from male to female. This in itself was not necessarily a positive development since the new work is exploitative of numerically flexible employees. As Walby (1989) has pointed out women who are employed on a casual basis in this way, are seen as a cheap, flexible source of labour.

Despite the production managers rhetoric, it seems unlikely that he, nor any of his colleagues were interested in positive discrimination in favour of women, but more that his views were consistent with the above. Women were a non unionized workforce and much cheaper to employ. Also they could be employed on a casual basis, releasing the company from many obligations that it would have to full time, permanent employees. At the same time it would make it possible to respond to peaks and troughs of business, and the companies would have the flexibility to deal with the next round of new technology.

There was generally more optimism for those who wished to redeploy at Mercury. Perhaps this was because the opportunities were likely to be greater due to the continuing expansion of the newspaper. There were opportunities for work as assistants helping in the planning of advertising, but the majority of opportunities were different posts in the production area, and the more demanding jobs in inputting in the various departments, as in Hermes. Mercury Newspapers had adopted a more positive attitude than Hermes, and instead of excluding workers, in most cases, they were generally given the opportunity to try the job and their success monitored over the year.

Technical criteria

Both of the newspapers were operating the same system of severance, paying those who wished to leave 'generous sums', and redeploying those who stayed. If they opted for the latter they would be paid the old compositor rate for one year without overtime. After one year they would revert to the rate of pay of the new job, if they were happy with the work, and the company was happy with their performance. If either party was dissatisfied the employee could then take another job at the rate prevailing for that job, or accept severance pay. The opportunities for transferring into different departments were optimistically predicted as being available in the advertising and editorial departments as well as production areas.

The companies were not prepared to operate the same policy as Zeus, and make people compulsorily redundant on the basis that the right ones do not always volunteer. However, there was an opportunity for selection of staff who wished to stay on and start new jobs. This was dealt with by the section manager interviewing the employee, who would inevitably have to get used to the idea of a lower weekly wage, and the loss of overtime payments which made up a significant part of a compositor's pay.

Had there been fewer volunteers for redundancy than were required, then Hermes would have reverted to compulsory redundancy on the basis of 'last in, first out.' Mercury Newspapers had declared a policy of not making any compulsory redundancies. The production manager at Mercury suggested the profitability of their newspapers could support a 'no compulsory redundancy' policy while Hermes could not.

Summary - newspaper studies

Confirming the hypothesis, considerable importance was placed on social and political criteria in both of the newspapers. Managers at Mercury echoed Cockburn's (1983) findings that the change in the nature of work, from a stereotypically masculine job, to a stereotypically feminine job, failed to bring a female staff initially. Managers use of women for inputting functions after direct entry technology is introduced, is about another form of gender typing, but this time women are viewed as a cheap, numerically flexible group (Walby, 1989).

The main allocation issue at both newspapers revolved around successful redeployment of potentially redundant NGA workers. Like Hermes, Mercury has experienced the same problems in the redeployment of NGA workers to editorial jobs. At Mercury this is despite intervention by the managing director.

The work of the editorial department was clearly seen as an attractive proposition by production staff for redeployment. It is an area where there has been redeployment in other cases of new technology introduction such as Portsmouth and Sunderland Newspapers (NGA, undated) and the Birmingham Post and Mail, where compositors were retrained as sub editors (Smith, 1988). Indeed, the greater demand for sub editors, and the similarities their jobs share with those of NGA readers, suggested that readers would have been ideal candidates for sub editorial training. However, both the newspapers were reluctant to use NGA workers claiming that they failed to reach the standards expected of trainee journalists. The NGA in both cases were unhappy about the rejections and suggested that the procedure was rather more conspiratorial than management had maintained. Exclusion by managers was based on several issues including traditional clashes between the editorial and production staff and the NUJ and NGA. Also the management were concerned about removing as many NGA workers from the company in an effort to minimize (or eliminate) the NGA presence.

The other manufacturing companies studied for their clerical jobs have something in common with the engineering studies in the relationship between age and perceived suitability for new technology related work. Similarities with both newspaper and metalworking studies extend to their attempts to divide the workforce between a casual group of workers and a multi-skilled group capable of responding to a wider range of tasks.

Office studies

The jobs in both Achilles and Apollo were in the accounts department and in production and distribution control areas. Age and sex seemed to figure prominently in the allocation criteria of both of these companies.

Social and political criteria

The general policy line had changed at Apollo who previously would take on many unqualified people and train them on the job. With greater demand for jobs, more qualified people were accepted for work in the company, few unqualified people would now get jobs. Also the change of emphasis to higher technology had meant that the existing technical staff had to be retrained, but there was a reluctance to train older staff. Early retirement for this group has operated as a means of removing the older workforce and replacing it with a ready trained workforce:

> ....it's been an encouragement to people who may have struggled on trying to learn new skills. But without being unkind to them, when you get to fifty eight or fifty nine it's not really worth the company spending a vast amount of money in retraining somebody. (Personnel Manager, Apollo).

However, the company's policy is to try and redeploy where possible. There have been a number of cases where staff have been taken from mechanical areas and retrained to work in computer software with the consequence that they go from the top of one hierarchy to the bottom of another. Although there is no loss of money, there are clearly likely to be problems associated with loss of status, although the personnel manager did not really see this as an important issue.

The financial accountant claimed there were no age barriers to retraining in the accounts department. He suggested that age was not important, but where new technology was concerned, an inquisitive nature and an ability and willingness to learn fast were the qualities required. As with the other studies the criteria for allocation of jobs was based on poorly defined concepts. Here the accountant used the term 'ability.' He maintained that those without such an ability, and unfamiliar with new technology would be at the top of any list of future redundancies.

Consistent with the newspaper studies, sexist ideas also figure prominently in the accountants allocation criteria. Invoice clearance, he maintained, was an area that required a high calibre person who was able to handle monotony. Women are suited to this work, it was claimed, because they do not mind monotony and are more interested in their family and children. He also claimed that the use of men in the section does have a 'stabilizing' effect on the women. This I interpreted as a further sexist assumption that with men present they talk less and work more.

In the manufacturing services division of the company a redundancy programme had removed 250 staff, and a policy of not replacing staff as they left had been adopted to avoid such an upheaval in future. However, this in itself had caused problems in peaks of business, and the division had begun to use sub contracted labour at these times. Once again the age issue emerged here. Some people, managers claimed, knew their limits and did not wish to retrain, while others were too old. The idea of age not being compatible with retraining was confirmed by the experiences of the operations manager, and so the policy of employing new staff had been adopted. When an attempt was made to retrain:

> ....they were the wrong age and it was a total failure; so there was a case where we tried to retrain and it failed. They just weren't capable it was as simple as that, we had to find them another job in that situation and bring someone else in, who was doing some other job, to do this new job created by the computer system (Operations Manager, Apollo).

Consistent with the other manufacturing companies, criteria for allocation at Apollo included both age and gender issues. There was evidence that Achilles were operating along similar lines.

Achilles had been through a rationalization programme, losing staff as a result of organizational changes and merger activity, but they were adamant that the loss was not due to new technology applications. There was some use of casual labour in the offices which, it was claimed, tended to employ young people who would eventually find jobs in the company anyway.

In the distribution area the use of casual staff was also commonplace: '....when you're sorting paper you'll bring in part time housewives who have got no qualifications at all....' (Distribution and Customer Manager, Achilles)

The department appeared to be attempting to build a strong skilled and flexible core and use casual work whenever required. With the implementation of new technology imminent, the firm began to employ older people as a workforce that would be easy to remove. The declared intention being that, when the new systems arrived, the older staff would be replaced by new young skilled staff. The view here was very much along the lines that the younger the people, the more amenable to change and to training they would be. However, the management information systems manager suggested that in fact that there had been considerable success in training older people, so perhaps the distribution departments ideas were poorly founded.

Technical criteria

The credit manager in Achilles ruled out any possibility of new staff being used in the accounts department. Any redeployment to new technology areas would be on the basis that those people were willing to accept change now and in the future, coupled with a degree of flexibility since the jobs were broadening. The financial accountant at Apollo also supported the view that new staff would be inappropriate because they would have no knowledge of the department or the methods employed.

APEX representatives expected any criteria to remove staff would probably be the traditional 'last in, first out' system on the basis that this was cheaper to achieve than removing the older staff. This would leave the remaining jobs for the older staff. However, the company's practice did not seem to be consistent with this. On the issue of age and allocation the union had given little consideration. It was often the case that changes were made without the union being informed by managers, and if the employees did not tell union representatives then there was no possible way that they could find out.

Summary - office studies

In Apollo gender and age seemed to have influenced allocation decisions alongside the tightening of technical requirements of staff. Also at Achilles sex and age influenced the policy for job allocation. In one department of the shoe company, both age and gender are exploited for their numerical flexibility.

The trend toward a dual workforce of skilled and unskilled workers, respectively a secure group and a casual group, was repeated in the allocation criteria of the companies. Further evidence of an exclusion of older workers on the basis of inflexibility and lack of ability was also present. However, here, as in Zeus, some important counter evidence existed to question the thesis that age was synonymous with inflexibility. Apollo had successfully trained older workers in computer operation of the accounts system. These members of the staff were working efficiently and reported no difficulties. Highly sexists ideas, that seemed to be questioning the integrity of female employees, were evident in the same department of Apollo and these were having a direct effect on allocation policy.

Unlike the newspaper studies, and some examples in the metalworking studies, it was existing staff rather than new staff that were favoured for the new technology positions in the office cases, because of their company knowledge and their established skills. Compared to the other studies there is less evidence here to support the claims of the hypothesis, but social and political criteria play an important role nevertheless.

Summary and conclusions

The first hypothesis of this chapter suggested that the answer to the question: what happens to jobs after new technology is introduced? - was not a simple matter of suggesting that deskilling or reskilling was occurring. In fact complex combinations of both of these were emerging. The analysis of interview data is presented under this section broken down into three groups of vertical reskilling, horizontal reskilling and deskilling. As expected different skill levels do exist in companies. However, evidence is strongest of horizontal reskilling.

Vertical reskilling refers to cases in which skills have been added to jobs at a level higher than workers have been accustomed to. Horizontal reskilling refers to the addition of an equivalent level of skills or the substitution of similar skills to jobs. Finally, deskilling refers to the simplifying of jobs, either through the removal of skills, or by the replacement of one set of tasks with another set of more simple tasks. This incorporates the Braverman (1974) definition of the separation of conception from execution.

Examples of vertical reskilling and deskilling tended to be sparse, and where they existed they were sometimes questionable. For the metalworking companies the transfer of programming tasks to operators was much discussed but unlikely to develop into more than the existing amending of programs (see Batstone *et. al.*, 1987). Comparable examples were in Achilles where there were plans to remove one layer of supervisory staff passing their responsibility to senior clerical staff. However, it became clear that much of the control function, a main function of a supervisors job, was to be handled directly by managers using new systems for surveillance purposes as occurred in studies by Jones and Rose (1986).

In the case of deskilling many routine jobs can be identified and could be described as Taylorist. However, the examples in the companies are generally not deskilled as a result of new technology but a similar division of labour has remained despite the technology. Hence, in cases such as the office and metalworking studies, work such as form filling or assembly may have changed in content with computerization, but the routine nature of the work is the same. In fact there may be a case to answer that this actually constitutes a reskilling of work, since the use of new technology constitutes a learning of new skills. Where deskilling is clearly evident is in the newspaper cases in which new technology has meant the simplification of compositors jobs, eventually resulting in their replacement with staff who were required only to have basic typing skills (see Cockburn, 1983 and chapter three).

The incidence of horizontal reskilling is by far the most prominent tendency in the studies. There were job expansions in Artemis where the training programme encouraged the constant learning of new skills, in Zeus maintenance workers were exchanging skills with each other. In both of the metalworking companies responsibility for quality control had been handed to individuals and taken out of the hands of a specialist team. Similarly, a range of computer applications in the clerical studies generally meant that more work was possible with fewer staff, and this inevitably resulted in the addition of tasks to existing workers jobs. Finally, reporters took on inputting functions that were previously in the hands of compositors.

In summary, there is little evidence of genuine vertical reskilling in the studies. In the cases where it appears to be taking place, a closer analysis generally reveals lack of commitment on the part of managers to follow this through. The same can generally be argued for deskilling, although this does not rule out the existence of these forms of organization in the firms. One reason that this may be the case is that managers recognize a need to exploit workers' skills particularly in cases of skill shortage and so will not wish to over simplify the work. On the other hand managers fear allowing too much worker control because the workers then have a much stronger bargaining position.

By comparison horizontal reskilling is far more common. The impetus for this is both new technology and other, more general, organizational changes. In some cases the tendency for new technology to blur job boundaries, as in the case of the electrical and maintenance workers at Zeus, means workers are given more tasks and skills. Elsewhere, such as in the addition of inspection work to jobs, these changes may be due to attempts by managers to create bigger jobs, reducing the number of workers in the process.

While the hypothesis is confirmed, the evidence for a mixture of outcomes emerging from new technology introduction is sparse. Although it is clear that there is no universal tendency for one form of work definition, horizontal reskilling is the most commonly occurring form of work organization after new technology introduction, if these cases are in any way representative. The Braverman *et. al.* deskilling thesis is not borne out by the data but neither is the vertical reskilling (flexible specialization) thesis of Piore and Sabel (1984) which has also been adopted in recent writings (Cavestro, 1989). The second hypothesis deals with the way workers are allocated to new jobs. New or different jobs are created when new technology is introduced because there is a difference in the kinds of tasks that workers are required to perform. There are a range of factors that the allocation process will be influenced by. The most obvious of these are technical qualifications which may include, skills, experience, academic and vocational qualifications. However, there are also social factors that may influence the way workers are allocated. Sex, race and age are three possibilities, and are factors that may ignore technical qualifications. Similarly, political criteria perhaps in terms of avoiding unionized employees are other possibilities that may take little notice of technical criteria.

Confirming the hypothesis there is ample evidence that demonstrates the importance of social and political influences in job allocation decisions.

Age formed important criteria in the four manufacturing companies since new technology was often seen as a medium that older people would have difficulty adapting to, this claim persisted even where older workers had already been trained successfully. Gender formed an important criteria in the newspapers and in the clerical studies for allocating workers. Like age, the perceptions about sex were conjured from extra organizational managerial beliefs. For the newspapers, women provided a cheap, highly flexible, and mostly non unionized group, all the virtues that they had not experienced in NGA employees (see Cockburn, 1983). Achilles were also using women on a casual flexible basis where contracts were short term.

Attempts by trade unions to influence allocation decisions were generally unsuccessful. The AEU at Zeus wanted allocation to higher grade jobs strictly on a seniority basis, but at the same time wanted to guarantee that no worker be embarrassed because they could not handle training. Managers insisted on a meritocratic system but allowed the no embarrassment clause. At Achilles, the shoe company, the trade union representatives were unprepared to tackle managers on allocation issues, while in Apollo, the Aerospace company, managers were unprepared to listen. APEX at Apollo had established a new technology agreement that had a 'no redundancy as a result of new technology' clause and promoted the redeployment of workers. However, the agreement was rarely adhered to by managers (see also Price, 1988; Cressey *et. al.*, 1988; chapter seven). The NGA had similar problems making the newspaper management listen, but had agreed that redeployment of redundant production workers would take place anywhere that there were vacancies. However, in the editorial departments of both newspapers managers prevented access on the grounds of test and interview results.

Criteria for allocation were not solely predicated on socio-political issues but involved technical decisions as well, although these seemed to be secondary. In both of the metalworking studies the technical ability and qualifications of workers was gauged, usually in discussions with supervisory staff and line managers, prior to new jobs being allocated or redundancies made. Zeus asked for volunteers and then assessed the ability of each of these, but even within this process, managers were interested in the attitude of the worker to the company. However, interestingly, age criteria did not pervade the maintenance areas where there was a high premium on multi-skilled staff. Artemis, however, were unprepared to relax their technical or social criteria despite concern over the shortage of skills in assembly and machining areas.

Since existing workers would have considerable skills and experience within a company, it could reasonably be expected that managers would choose redeployment options rather than employing new people. This was the case in the accounts department at Apollo where redeployment and retraining were favoured above taking on new staff. In contrast, the assembly department at Artemis had removed semi-skilled fitters and testers and replaced them with new staff who were skilled fitter/testers. Similarly, in the newspapers, where production workers applied for editorial jobs, managers decided against their employment on political grounds. In these cases, socio-political criteria become more important in decisions of allocation than technical qualifications.

In most cases social and political criteria were not only employed alongside technical considerations in decisions about allocation, but as criteria capable of overriding technical decisions. Particularly good examples of this are in the newspapers. Hence the hypothesis, that political and social criteria will predominate in allocation decisions is accepted. Evidence suggests socio-political criteria may acquire greater importance than technical criteria.

The next chapter considers, for the six case studies, how the new technology is introduced, why it is introduced and who is able to influence the process.

Notes

1. This was largely to deal with the shortage of skills that Artemis was experiencing. See also chapter two, section on flexibility.
2. 'Copy' refers to anything written to be printed in the newspaper - stories, advertisements, sports results etc.
3. It was a recurring theme in the interviews that the errors in stories were not the fault of journalists but of compositors who mistyped them. The same argument is applied to contributed copy. However, during a visit to Mercury, in the early stages of direct entry usage, a major error in the form of a spelling mistake in a title on the front page appeared on the main evening newspaper. It was discovered at a time when the production process was too advanced to be stopped.
4. The computers used in direct entry photocomposition have queuing systems that allow reporters to type in their text which is passed to a sub editors queue where it waits until the sub editor has a chance to call it up, read and edit it.
5. See chapter two.
6. This was a quotation from the manufacturing manager at the company.
7. Problems with unskilled staff were reported in some newspapers by NGA FOC's and the recall of skilled workers was the result. See Rainbird (1988) and the debate on tacit skills in chapter two. One Welsh newspaper was identified as having this specific problem - the South Wales Argus.
8. See also Cockburn (1984).

6 Managing the changes in work organization

Hypothesis

In the work organization aspects of programmes for the introduction of new technology, formally planned strategies for change are rare. More often changes are dealt with on an *ad hoc* basis as and when problems arise. Rationales for the use of new technology are varied and do not reflect any general motive of managerial control over the workforce.

There is an inextricable link between planning and the actors involved in this process. The latter part of this hypothesis maintains that involvement in the planning of work organization is confined to senior management levels and rarely reaches the workers and users.

Introduction

Following on from consideration of the nature of work, and the allocation of work after new technology is introduced, this chapter considers the decision making process and the justifications for new technology. The hypothesis is divided between three issues: strategy, rationales and discussion for change. These three issues are closely inter related.

Strategy often forms the basis of change. The presence of strategy suggests that managers set aims that a strategy is directed to achieve. Aims without an established way of achieving them leaves the outcome in question. In respect of work organization, a lack of strategy, and a tendency to deal with problems on an *ad hoc* basis, may result in an untidy and inefficient use of human resources. Lack of definite strategy is a finding of Rose and Jones (1984), Hyman (1988) and Burnes (1989; also Child, 1972). Braverman (1974) has suggested not only that strategy exists, but a general tendency of deskilling to achieve managerial control is evident. Child (1984) on the other hand sees strategy for deskilling and control as secondary to other motives such as efficiency, and Salaman (1986) has argued along similar lines.

Rationales or justifications for introducing new technology may not be guided by any strategy. Rationales are different from aims. An aim is an intention of what is to happen, but a rationale is often applied concurrently or retrospectively to explain why a particular development occurs. Rationales may relate to aims but need not do so directly. It is possible to have a very wide range of justifications for particular changes which vary according to departments and individuals, and have occurred independently of aims. Wide variation of rationales may also be due to a lack of strategic planning because neglected issues have to be picked up elsewhere at a later time. Hence, an absence of corporate planning leaves considerable room for line managers to interpret and implement their own rationales for introducing new technology.

Strategic planning, aims, and some rationales in new technology programmes, are often informed by consultation that takes place prior to the making of policy or the implementation of ideas. If this consultation, in respect of work organization, is confined to managerial staff, essential knowledge of jobs will be neglected. Cressey (1990), for example, argues that worker involvement is rarely prominent at the policy making stage but is more common later in the implementation process. A result of this may be that work organization and specific job definitions may allocate tasks inappropriately, and possibly, fail to include some essential tasks, leading to further incremental changes, and creating further problems of work allocation.

The case study evidence confirms the hypothesis, strategy for job design is absent in most cases. Rationales are varied, but do suggest motives of managerial control over workers in some cases. While influenced by various actors, the change process rarely includes worker consultation. Where workers are involved, any discussion is often confined to advising staff and not listening to their views.

Managerial strategy for introducing new technology

At the root of all decisions to introduce new technology there must be some form of planning, even if this is only to decide what to buy. Key decisions are, considerations of expenditure, and the purpose to which the new technology is to be put. These discussions need to be conducted at a range of levels, not merely at board meetings, where funding is allocated. Besides the technical considerations, there are also social ones that involve issues of job and work organization changes. Lack of prior consideration of these factors may not only result in inappropriate choice, but also inappropriate deployment of equipment.

Metalworking studies

Artemis, producers of hydraulic systems for aircraft, had technical reasons for their use of new technology. The precision nature of the products they produced meant that very long lead times would be involved. CNC machinery was capable of speeding up the process because more machining could be done on one machine, and less time was spent in transferring the component between machines.

Both the manufacturing manager, and production director at Artemis, argued they had spent a good deal of time in planning and preparing a strategy for the future, one result of this is the grading structure. There seems, however, to have been a much greater investment in the planning of the technical aspects than the jobs themselves. The machine shop changes were considered five years previously and involved details of machine tools, the general effects on the workforce in terms of numbers, and back up requirements such as production engineers. The importance of job planning was played down since CNC had existed for some years and little change was envisaged. However, the machining centres arriving in the factory were quite different to anything the company already had, and the concepts of multi-manning and unmanned running were unknown.

One area where evidence of strategy for job design was offered, was in the form of the module training package that the company devised to improve flexibility. This also formed a means of progressively upgrading the skilled workforce. Nevertheless, while workers were consulted by the superintendent to establish where they were in respect of newly conceived job gradings, and the level of module programmes to be aimed at, it was felt by the technicians themselves that the jobs were not planned but '....just developed, that is the only way to describe it really, as they've got more technology in, then everybody's grown with it.'

It is apparent from this statement that the process, for workers, was more one of adding bits to different jobs as the new technology demanded it than could be described as strategy. Consistent with work by Burnes (1989) and Hyman (1988), planning was limited particularly as far as the details of new task structures were concerned. Managers may claim that shopfloor perceptions were inaccurate, but by their own admission planning for CNC jobs had been limited.

Surprisingly, the senior shop steward for the AEU, working in the machine shop, suggested there was careful planning of jobs at which trade unions were involved and able to influence policy. The lack of any examples where the trade union had changed the direction of policy, and his general lack of awareness of fundamental issues affecting the workforce, led me to doubt this was the case. He even presented contradictory evidence for his claim where information for the arrival of new machines was never longer than three months in advance. However, as in studies by Davies (1984) it is possible that this level of information provision was acceptable. No form of consultation or negotiation took place. The senior shop steward did suggest that there was a procedure of asking workers what their experiences and preferences were prior to buying machines. However, this seemed to be more a desire to find some evidence consistent with my question than something that was a central indicator of consultation. Although it is clear that the overall flexibility ideal was planned and discussed with the trade union, it is far from clear that the 'micro' job design issues, in terms of the tasks workers would perform, were involved.

Managers and workers contradict themselves on the existence of strategic planning at Artemis. However, in areas such as job design, managers sometimes failed to see the need to plan. Whilst planning may have existed in the choice of technology system this was based on technical criteria and neglected job design issues. At Zeus, consideration of job design appeared to be more concerned with avoiding the concession of grade increases.

Zeus, the diesel engine manufacturer, wanted to reduce their costs. A depressed market for diesel engines highlighted over staffing, and new technology introduction provided an opportunity for reducing staff numbers by making production more efficient, and merging the remaining jobs.

One of the key production operations managers at Zeus explained that the company had not always planned so carefully. However, more careful planning was prompted when a grade was conceded on new computerized test equipment that was introduced. From this point a system of thinking ahead and adding tasks to jobs was devised, even though these functions had not yet been established in the firm:

> ....we've tried to think several years ahead as far as the bounds of technology of saying you will use strip readers and bar codes and we've added those things in, even though we didn't have them.... And what we're saying is we've got 500 of your colleagues here that want to volunteer to be redundant and we are prepared to let them go, but to be able to let them go we must change our structure, halve our inspection department, do away with our progress department and you must take on board some of their skills.[1]

Part of this involves appealing to the workers moral duty to help their colleagues. As in Artemis, there were also increased rewards on offer, which seems common in cases of applying new technology (see Batstone *et. al.*, 1987). Hence, the workforce gets a higher grade but they must operate VDUs, access the mainframe, use bar code readers, write out labels and so it goes on. In terms of controlling grading, reliable methods have been developed.[2]

In other areas, the formulation of a strategy that dictated a certain financial commitment, had not been followed through precisely because of the financial costs. One production engineer described it as the company making a major investment in machinery and then 'pulling back on the last ten bob.' The machinery in this case was CNC, and the original commitment was to develop the operators' programming skills. The measures only went part of the way and the result is a gap in the knowledge of operators and unfulfilled expectations. This generated a degree of bitterness amongst the CNC operators interviewed. Perhaps worse than having no strategy at all is having a strategy that is abandoned before it has had a chance to develop (Jones and Scott, 1987). The irony is that the technological investment is considerable and the full benefit is not being obtained from the machines because the commitment to train the operators has not been fulfilled. All the emphasis is once again placed on the machine and not enough on the operator and the jobs.

Newspaper studies

At both of the newspaper companies, the aims of introducing new technology were to remain buoyant in an increasingly competitive market. New technology was being adopted in regional newspapers nationally, but local companies were also using new systems to produce free newspapers. There was a general feeling of the need to reduce costs, and improve quality, to make the newspapers more attractive to advertisers and readers.

Contrasting with the metalworking studies, there is more evidence of a plan both to introduce new technology and to reorganize jobs in the provincial newspaper industry. Managers of Mercury Newspapers (the larger of the two newspapers) put forward the same arguments as each other when asked about strategic planning for technology and for jobs. There was no difficulty in deciding how technology was to be deployed nor how tasks would be distributed. Everything, it is claimed, is set by the technology which is fairly standard throughout the industry.[3] However, the NGA father of chapel expressed his doubts on the eve of implementation: '....I get the distinct impression it is still being thought about. Partly because they really didn't foresee the problems....'

Clearly technologically deterministic approaches cannot effectively predict the social requirements of work design and this was particularly true in the case of the newspapers where new technology deployment was itself political (see Wilkinson, 1983). The management had apparently decided to take on production assistants to undertake the tasks of inputting copy

which, it was assumed, were easy to perform. However, they failed
to consider the complexity of preparing advertisements:

> I really believe in my mind that the company haven't got a clue, I think they just
> thought they'd bring this thing in and as long as they got a 'paper out every day it
> wouldn't really matter how much it cost for a period of six months or eight months and
> they would really haphazardly see at the end of that six months who they needed and who
> they didn't (NGA Father of Chapel).

This suggests that tacit skills, that some writers claim
protect workers against deskilling (Kusterer, 1978) and job loss
(Rainbird, 1988), may become worth little in the face of a
determined management, who may also be indifferent to the details
of task execution.

There was more evidence of what seemed to the workforce to be
an *ad hoc* approach at Mercury, in the failure of the management
to adhere to agreed levels of voluntary redundancy. Here the
company decided to change course and let anybody go who
volunteered for redundancy or early retirement.[4] The unions
reflected on the unethical nature of the break in the joint
agreement but found themselves in a dilemma since the removal of
more staff opened more opportunities for those remaining. The
combination of unplanned situations led the FOC interviewed to
conclude that it was by no means a planned approach but one of
'progressive evolution.'[5]

On the editorial side of the business the NUJ FOC conceded that
it is not possible to see all the problems in any system, so
there are considerable limitations to planning. Essential details
such as the use of portable computers and modems by journalists,
had not even been discussed and were being left to the time when
they were actually required. The assistant editor also suggested
that planning had been limited because many problems were arising
in the system and were being 'ironed out on a continuing basis.'

Mercury depended on new technology for direction in change to
work organization. However, the managers seemed to assume that
new inputting staff would be capable of taking over where
compositors left off, while the trade unions were concerned that
inputers would not be familiar with procedures. Hermes
experienced similar problems with staff deployment and had
avoided consultation with the production department.

Hermes had acquired a strong 'editorial bias' (the editorial
department had the upper hand in decision making between these
key departments), stronger than in Mercury where, nevertheless,
such a bias already existed. This was important since the
newspaper appeared to start out with a 'production bias' and the
change occurred when the original managing director left the
firm. At that time the production manager was relieved of his
responsibility as 'direct entry coordinator', which was then
passed to the deputy editor. Apparently from this point the
coordination and planning of the system considered the editorial
side of the business rather more than the production side. In
both newspapers work organization began to resemble the Edwards'

(1978) model, with a primary workforce of journalists and editorial staff, and a secondary workforce of inputers and production staff. However, as in Mercury, the problem with such an approach overlooked the importance of the skilled staff remaining in the production area.

The direct entry coordinator showed little interest in the production side. He explained that the approach was to visit a comparable regional newspaper that had introduced the technology and glean knowledge of training periods required for editorial staff. Cockburn (1983) and Smith and Quinlan (1982) refer to the case of the Croydon Advertiser in which new technology was seen as a means of fusing the editorial and production staff who had traditionally been two sharply defined groups. Both Hermes and Mercury, influenced by the strength of the editors, appeared to be seeking to achieve the opposite. That is to maintain a division establishing the editorial staff as a primary workforce distinct by the nature of the work they do and their level of pay.

Office studies

Amongst office jobs, in the two other manufacturing organizations, similar mixes of strategic planning and *ad hoc* implementation styles were found. Achilles, the shoe company, was introducing new technology in several of its departments. The same was true of Apollo, the aerospace company. Neither firm had established plans for new technology on a corporate basis but there were a mixture of departmental aims. A key one of these was to improve information flow, opening new possibilities for the dissemination, and presentation, of information for the benefit of customers, and in the interests of departmental interdependency.

Achilles, the shoe firm, had appointed a manager to specifically look after the developments of a new electronic point of sale system. This had been a very carefully planned scheme in which a pilot study had been set up in some fifteen shops around the country. The system clearly involves a considerable capital investment and the purpose of the pilot studies was to justify whether such high costs were warranted. The view was taken that the jobs were not really changing but the tools to do them with were, therefore, discussing the jobs was of little importance. In fact in reality there were considerable qualitative shifts in the job structures.

Similarly, the distribution and customer manager at Achilles reported that the trends the department had experienced in job enlargement, staff reductions and higher average gradings were all unplanned aspects that occurred as a result of new technology. The aim to achieve automation of the factory and increase the efficiency with which the stock was moved inevitably involved reducing the paper work load that the clerical staff were having to carry. Obviously the role of the computer systems would be to take over much of the paper work. But it was claimed that management do consider the restructuring of jobs carefully, if some or all of the functions are to be computerized. Indeed, the process of discussing jobs has developed more and more as the computer applications have increased.

However, it was conceded that the department found it very difficult to plan ahead and define jobs with any accuracy. Consequently, Achilles had adopted much greater flexibility within its workforce, on the basis that it could then use staff in one job to look after any other new tasks that might emerge and that were previously unforeseen. Here, as in the newspaper studies, two groups of workers seemed to be forming, consistent with Edwards' (1981) primary and secondary groups of employees. The nature of the contracts involved also suggests changes consistent with Atkinson's (1985) model. In the transition period to new technology Achilles took on many older staff and female employees on short term contracts. They considered these dispensable, this group looked after the more menial manual functions in the distribution department. When the new system arrived these staff were removed and the permanent multi-skilled staff, the indispensable core, took up their new responsibilities. The fact that the multi-skilled group was flexible across a range of tasks, meant, for the managers, that planning precise job definitions was unimportant because these people could be moved from one job to another freely, or have additional tasks added to their work when required.

Once again, in the accounts department at Achilles, the credit manager explained that there had been significant discussion regarding which system to buy. But it was clear they had failed to consider the task structures at the same time.[6] However, there was a view that the system would allow divisions between the staff. Having one group of skilled staff handling the more complicated work and another of numerically flexible staff. These rather superficial decisions were arrived at, on the basis that a new integrated system would allow a free flow of information, and less supervision of staff would be required as a result. None of these changes had been discussed with staff or trade unions, and there was no intention to do so. Hence there was a strong likelihood of problems occurring since the plans involved the removal of some supervisors and deskilling of others against a background of a reskilled group of key staff.

Achilles' managers gave considerable thought to the technology changes themselves, but paid insufficient attention to the jobs. In one department the use of a highly flexible group solved the problem of deciding on precise job definitions. Other decisions about the changes to jobs involved a radical transformation of the department that was highly theoretical and practically implausible. At Apollo, managers saw little choice in the way jobs were designed.

Interviews at Apollo revealed very little planning in the personnel department, and on one particular aspect of new technology (word processors) the personnel manager commented: 'There is not, that I can discern, a site plan for the introduction of word processors, it is very much an *ad hoc* departmental thing.'

The personnel department had computerized their databases, which was information previously held on cards, using a basic computer package on a personal computer. But like the unplanned wordprocessor technology the personnel manager pointed out that, '....we used that almost by accident.'

In the accounts department at Apollo a separate section had been created to consider potential new computer applications. The financial accountant here saw little choice in work organization however, which was consequently poorly planned. In the payroll section, set aside from the main department, the staff had organized their own division of labour and methods of working. The payroll section was managed by a supervisor, who was given considerable authority because the section was physically isolated from the main accounts department. The division of labour was based on the supervisor's conception of being fair and not leaving one group of staff with varied jobs and another with detailed boring work. The work was shared between personnel equally and not according to grade.

However, on the manufacturing side of the business there was a much more clearly defined process of planning derived from pressures to keep stocks low. The introduction of new technology in these other departments, necessarily, has to be planned and coordinated carefully because there is a direct link with other departments. Also here there was a structured organization of how the jobs would change in the form of a job grading committee who decided on the levels the staff should be in the organization. Job descriptions are drawn up by departmental managers and agreed by the employees.

The operations manager of the management services group at Apollo talked of a random system of finding the right machine first, and then adapting the jobs to meet its requirements. Hence the reallocation of workers and training always takes place after the introduction of new systems. Job specifications themselves were often set by finding someone who was not fully occupied and asking him/her to take on a new function. However, the operations manager claimed that while there was considerable planning in theory, this failed to relate to the way the system turned out in practice. People could not cope and there were a range of *ad hoc* responses in amending the system:

> .....I think it's just a salutary lesson that you can think you've done everything, you understand the system but the real telling point is when you put it into action that's when you really know and all you can do at the systems development stage is try and knock out ninety five per cent of the likely bugs, there will always be one that will catch you.... (Operations Manager, Apollo).

Belief that there is little choice in the way jobs are designed, and definitions are dependent on the new technology that is introduced, meant that work organization in some areas was poor. This was true of the accounts department. Elsewhere, the importance of job planning had been considered but planning had not been able to prevent later *ad hoc* changes. Apollo demonstrates that even where job aspects are considered, outcomes may not resemble expectations, and as a result job definitions may be inadequate.

Summary - managerial strategy

In both newspaper companies managers argued that the organization of jobs was not an area of contention since, at least as far as the managers were concerned, the technology itself set very clear job definitions. However, many of the workers interviewed felt that there was an *ad hoc* approach to the technology and the jobs. Examples were cited of a lack of training on the editorial side and a failure to consider key posts on the production side. Nevertheless, both companies seemed to be adopting the same overall approach to work organization, to segregate the two groups of editorial and production workers and assert the superiority of editorial staff. This also involved a major reduction in the skill levels of the production departments.

At Artemis there is an apparent lack of strategic planning for the organization of jobs after new technology introduction, although where the capital equipment is concerned, more careful consideration is given. Zeus has attempted to link its planning of new technology with the design of jobs but this has been conducted in an instrumental way. In respect of jobs, planning has avoided regular grade increases by including new tasks in job descriptions, even before they appear. This process involves advanced consideration of the new technologies to be introduced and the tasks required to operate them which constitutes some form of strategy.

Confirmation that Zeus was following an instrumental approach, and not one of the kind described by sociotechnical theorists, can be found in evidence where conventional machine operators were redeployed to take on CNC functions. Due to cost constraints the company failed to follow through the concomitants of job training and additional pay.

In the accounts departments in both of the other manufacturing companies, the managers failed to see the potential variety in work organization. Instead they tended to assume a work design already set by the technology in a similar way to the newspaper studies. In the manufacturing services division of Apollo the manager admitted inadequate planning in the introduction of new technology first and then fitting an employee around the tasks. In common with the newspaper studies, this is a style that implicitly assumes that the technology defines the jobs. In the distribution area of the Achilles' organization, there was evidence of a dual labour market strategy of functionally flexible and numerically flexible work establishing two stratified groups of worker. Furthermore, the functionally flexible group was receiving multi-skill instruction, which relieved the Achilles management from deciding the precise definition of jobs. Not only did it release them from the concern of work design, but it also meant that decisions about the allocation of workers were less problematic.

Closely linked to the planning, that goes into deciding to use a particular system or a particular job design, is the very reason why the new technology should be introduced in the first place. If these reasons involve the workforce characteristics or problems, it is a reasonable assumption that they will have a direct influence on the actual definition of jobs. Hence, it may be possible to identify certain styles of job definition according to the rationales that managers apply to the introduction of new technology. For example, a rationale of deskilling may be linked to a strategy for achieving control which is manifest in a detailed division of labour. These issues are considered in the next section.

Rationales for new technology introduction

Here, rationales are taken to mean justifications for introducing new technology. Rationales differ from aims since aims are prospective and refer to intentions. It is expected, that unlike aims, rationales will be numerous and quite different according to the perspective of different actors. They will also depend on the extent to which strategic planning has taken place in companies. Where there are gaps in planning, rationales are often invented by line managers to explain changes they wish to make which are attributable to new technology.

Metalworking studies

Rationales for new technology at Zeus were identified by production managers as improving productivity and quality. The industrial relations manager saw them as a means of reducing staff levels and increasing the accountability of staff. Artemis identified a series of rationales close to the organizational aims. Overwhelmingly, economic justifications were prominent but often defined in terms of improving competitiveness through shorter lead times, quality and service.

New technology was not expected to provide improved productivity and quality at Zeus in the short term, but it enabled them to come more easily over time. Technology was also used, more specifically, as a medium to achieve an alternative organization of jobs. For example, aided by computers, the stores are able to see if stock is running low, and can then search to see if there is an alternative stock elsewhere, thereby ensuring that the stock is maintained. Before, it would have necessitated a trip to the stores to inspect and physically establish the stock levels.

It was further argued that technology provides better quality products in the case of CNC, and better customer service, where the computer is now able to locate orders in the factory, so the customer can be kept informed.[7]

Conversely the industrial relations manager at Zeus justified new technology in terms of a need to reduce the number of employees in the factory, other changes in jobs were also aimed at this idea, and were not necessarily the result of changes in technology. For example, amalgamation of the setting and operating functions.

In the case of the maintenance function the job change was directly as a result of the technology, and particularly of CNC machining. As Rainbird (1988) has pointed out, new technology introduces particular problems in maintaining traditional demarcations. Training proved inadequate amongst the fitters and electricians because new equipment demanded knowledge of both trades. The maintenance supervisor explained: 'the way machines are built it's a problem sometimes to say whether it's mechanical or electrical, or whether it's hydraulic....'

The resulting problem was that fitters were always waiting for electricians, and electricians for fitters. The equipment that was being introduced also dictated a need to take electricians more into the field of electronics. So training was instituted to teach electricians mechanical skills and fitters electrical skills so that each could do eighty per cent of the others job. This now means that one person can take responsibility for a complete job.[8]

Zeus has also pursued expansions in the task range of jobs based on new technology and justified them because they allowed cost cutting. The removal of quality inspectors saved money and, arguably, it improved quality, although the workers on the shopfloor were at odds with the latter view. At Zeus the personnel manager raised accountability as another rationale. Machine operators, including CNC workers, must be accountable for the work they produce. Some operators saw this as providing some pleasant spin offs, in terms of greater responsibility, that served to enhance their jobs, others were more sceptical. However, contained within such an approach there is a potential element of surveillance, since managers required the workforce to check the work at each stage and confirm that it passed quality standards by certification with personalized rubber stamps. If a component contained faults, the individual responsible could be identified easily. Here, managerial control motives may actually underlie the apparent enhancement of work (see Pignon and Querzola, 1976).

Zeus identified a range of justifications for the introduction of new technology. These range from technical reasons of quality, economic reasons of increasing productivity and reducing the numbers of workers employed, and control issues of accountability. The rationales given often deviate from the aims of the company but they reflect the specific interests of the respondents. Managers at Artemis claimed that changes to jobs were unrelated to the introductions of new technology in themselves.

At Artemis, the reluctance of both the manufacturing manager and the production director to concede that the job changes had in any way been a function of the new technology, contrasted sharply with the opinion of the machine shop superintendent and the senior shop steward.[9] The personnel manager saw new technology as a catalyst in job changes, he saw new technology and job changes as inseparable.

The machine shop superintendent went on to argue that the rationale for new technology was mainly the reduction of long lead times which were two or three years but now are down to just one year. In common with studies by Jones (1988) and Burnes (1989) other important factors he referred to were staying competitive in terms of quality and price, but also less maintenance and high quality were combined with reliable repeatability on CNC machines. The personnel manager agreed but added that one production unit which had provided 'a long line of bread and butter' was now required less and, as a result, the factory had to be tooled up to accept a more diversified product range. Here there seems to be a shift away from a Fordist design structure in that new technology is employed to increase flexibility to handle more variable product groups (Sabel, 1982).

Economic justifications for new technology introduction were prominent at Artemis. The personnel manager claimed that without new technology the competitive position of the company would be threatened. However, the industrial engineering manager was less prepared to assume that benefits would necessarily be produced by the technology: '....unless you can lower costs overall there's not much point in introducing it. That encompasses improving quality and reducing lead times' (Industrial Engineering Manager). Here there are a series of rationales for new technology introduction; precursors to an ultimate goal of lower costs.

Much of the new technology development in the new build assembly area, and in repair and overhaul, seemed based on keeping the customer satisfied. The computerized test and pre-test units employed were purely to simulate the kinds of test that customers would put the hydraulic units through. The need for the equipment arose because customers were using more stringent testing equipment than Artemis and rejecting units as a result. However, the technological introductions in both new build assembly, and in repair and overhaul, were minor and the repair and overhaul manager denied that technology had any significant effect. Instead he emphasized the policy to increase the flexibility of the workers in accordance with other job changes. The two departments provide an excellent example of this flexibility in their ability to exchange staff when the work load dictates a need.

On the one hand there was a reluctance to attribute job changes at Artemis to new technology, on the other these were seen as inseparable. On the whole new technology rationales were close to the aims of the organization. Other rationales in respect of CNC reflect the findings of other researchers (Jones, 1982; Burnes, 1989).

Seeing new working practices as a necessary part of dealing with new technology has parallels in the clerical and newspapers' studies. The newspapers have a number of aspects in common with the metalworking case studies in the emphasis of improving productivity to remain competitive, but much stronger under currents of a managerial control motive were also present.

Newspaper studies

At both newspapers, new technology introduction was justified in terms of a need to update old, and difficult to obtain equipment, a problem recognized by Smith and Quinlan (1982). However, linked to the use of the new equipment was the recognition that it was possible to eliminate the typing of work twice, meaning greater efficiency.

For Mercury and Hermes Newspapers the 'archaic nature' of the production of newspapers within the organization and within the industry as a whole was a main concern. The production manager argued that the urgency of change now was prompted by the increasing threat from local television and radio, an argument well aired by the Newspaper Society but challenged by the NGA (NGA, 1983). Smith (1988) has also noted that the cost of producing newspapers has become relatively low with the availability of new technology and the result of this competition is inevitably that the local newspaper market itself is threatened.[10] There was also a strong element of control involved, this was not explicit and was actually denied but statements were made pointing to the necessity that the union should be broken and new technology would achieve this (as in Zimbalist, 1979b). For example: '.....it's only the unions that held the thing back.....I mean to put it crudely they got away with it because there was no competition at the time and you would screw your customer' (Production Manager, Mercury Newspapers).

The change of climate has meant that advertisers can now go elsewhere, so it is no longer realistic to think that the customer can be required to pay more. Newspapers now have to operate with the new technology or face the real risk of bankruptcy.[11] The increase in the quality of the production of newspapers has also been a driving force including the increasing use of colour, which has in general been added because of the greater demands from advertisers. However, the hot metal production methods were unhealthy and unsafe. The main dangers were from lead oxide, molten metal and the use of chemicals such as paraffin.

For the assistant editor at Mercury the new technology systems meant flexibility and a significant part of that flexibility was the destruction of traditional trade union demarcations. The unions' rigidity prevented shift working and expansion of the newspaper. But the newspaper was also aware that NGA staff are highly skilled with training of seven year apprenticeships and wages to complement this, but performing jobs that only require basic skills such as those of using scalpel and paste and being able to key in text. The rationale here was to reduce costs by employing cheaper workers.[12] A significant influx of staff who were trained 'on the job' in a very short time, followed the implementation of direct entry technology. The message was once again of a changed climate and the need to remain competitive within that climate:

> ....if you don't have the advantages of direct input, you really can't run a competitive newspaper....it's just impossible to run a newspaper which is competitive under the old rules and I think over the next ten years, unless papers do recognize this and go over to direct input, many of them will go out of business (Assistant Editor, Mercury Newspapers).

Greater efficiency to maintain competitive position is clearly uppermost in Mercury, but there is also a recurrent theme of removing trade union control that stood in the way of direct entry technology. This suggests that to achieve the aim of greater labour and capital efficiency the NGA power base had first to be dismantled. Hermes did not differ greatly from this scenario and like Mercury, Hermes was always concerned about the quality of the staff in the composing room and other NGA controlled areas.

The NGA operated a pre-entry closed shop, that is, a person had to be an NGA member prior to finding employment as a skilled production worker. The NGA held books of eligible members who were not always the people the management wanted to take on. Once again, here the new technology proved to be a means to an end in achieving the breaking of not just the pre-entry closed shop, but also any sort of closed shop that the NGA or any other combination of unions would like to have maintained. As in Mercury, control motives are necessary to ensure union control is displaced, before the use of the new systems can be fully established.

The reproduction overseer suggested that the rationale for change at Hermes was that the equipment needed replacing and few suppliers were maintaining the old machinery as a direct result of this (Cockburn, 1983; Smith and Quinlan, 1982). In his view the quality produced with photocomposition systems was not as good and the claim of cost saving was no more viable because there are so many consumables. Film and paper are more expensive than lead since lead was just one of the materials that could be reused, only requiring occasional topping up. Also the zinc used for etching photographs itself had a resale value. However, the work is cleaner, safer and the staff are cheaper. The other dimension was the de-unionization, although the new staff are not prevented or discouraged from joining a union, he suggested that it was unlikely that they would because they were women and probably not the main wage earners in the family, therefore, the jobs are likely to have a very high turnover.[13]

For the deputy editor at Hermes the rationale was much more straight forward:

> ....technological changes have happened....working a hot metal linotype machine was a more skilled job than working a computer keyboard but it was, in terms of the total number of lines you could produce, much less efficient, simply because the computer can do things for you that previously involved your skill and decision making when you were working on linotype.

Justification is therefore based on the much greater labour and capital efficiency and particularly dependent on the elimination of 'double keystroking.' The same arguments as those in Mercury were put forward but there was particular annoyance at the high costs required to maintain NGA workers. Not only were the NGA people often paid more than advertising and editorial staff, but they would be very expensive to employ at overtime rates so that special editions of the newspaper were out of the question. There was general resentment about restrictions imposed by union control:

> (a) We know the traditional composing room restrains what we can do. (b) It's very expensive to continue and (c) it's totally pointless to type out the story twice. And then the decision becomes obvious; what we need is a direct entry system.... (Deputy Editor, Hermes - his emphasis).

Here again direct entry technology introduces the motive of dismantling union craft control. So the introduction of direct entry also increases flexibility. However, the deputy editor at Hermes was typical of many of the others interviewed in suggesting that this was neither an aim or a justification of new technology but was a precursor to its introduction. The real reasons he claimed, were the capability to respond to the market more fluently and compete without the high production costs and the restraints that were experienced with traditional labour (Martin, 1981). This is equivalent to central questions of flexibility and of being able to circumvent the union power base to achieve such flexibility. A view shared by the Hermes NUJ FOC: 'I'm sure that the primary intention is not to smash any particular organization, but a happy consequence of the new technology is that it enables management to change the working relationships within a company.'

According to the NGA FOC, the main aims of direct entry technology were increases in terms of labour and capital efficiency. It also meant reduced costs for the newspaper, but perhaps more important for the political dimension, it means that the NUJ and NGA no longer have to work together in close collaboration. This, in the past, had offered the possibility of a range of problems from union convergence to the greater divergence of the two unions.

Control over workers is also a significant feature in the rationales of one company in the office studies. Productivity and quality were as prominent here as in the newspapers and the metalworking companies.

Office studies

Avoiding an increase in staff, increasing worker flexibility, and customer service were referred to as rationales by various managers at Achilles. However, in the absence of an overall company plan for new technology, departments were left to define their aims. This means that aims and rationales are closely associated. Apollo, quite different from Achilles in the nature of the products produced, was concerned with integrating computer systems to speed up the administrative processes and flow of information. Rationales here often incorporated motives of managerial control over workers.

The systems manager at Achilles reported that new technology was used as a means of avoiding a heavy and expensive influx of staff, which emerged as the only alternative to enable the more rapid distribution of shoes to the Achilles' owned shoe shops. The problem was not reluctance to employ the staff alone, but was also one of finding suitable staff to fill the positions. The suggestion here is a use of technology to alleviate skill shortages as in Jones (1982), and a merging of jobs to deal with the shortage and respond to the new systems. There are useful comparisons here with the Artemis Engineering studies where multi-skilling of workers seem to be a response to a market shortage of skilled workers. But new technology systems were also capable of realizing the aim of providing a better quality of service to customers and providing a fast turnaround of stock in the shops.

The personnel manager at Achilles saw flexibility as of key importance in his department since there was a heavy reliance on a few key people. Inevitably if one person was sick then the information required was difficult to find. A computer system enabled a general reduction of staff and less reliance on specific people. There is now more access to information and the work has become less routine. However, this also suggests reduced autonomy, possibly meaning less individual control and job security, although these were not mentioned as rationales for new technology usage.

For the distribution and customer manager, new technology had been introduced in the warehouse to improve its productivity and efficiency. However, retrospective justifications for new technology included improving customer service. At the same time the possibility of a hidden agenda was strongly evident, staff levels had halved and the output of shoes had more than doubled. However, there can be little doubt that with so many customers, in the form of retail shops, all calling for faster response times, there was strong impetus for an improvement that also allowed them to operate with relatively small stocks. The management information systems manager explained, that the new system also means rapid and easy collation of data, enabled by progressively providing a more varied choice of selection parameters to facilitate easier, and more rapid retrieval of specific information.

In the accounts department at Achilles the credit manager suggested that one rationale of adopting new technology was to present the right image to the customer and if the company and department were to retain a high quality of staff then it was necessary that the right tools were offered for them to use and make decisions more effectively. This particularly involves improved access to information, made easier by integrated computing facilities. The manager explained that debt control had been maintained effectively with a system of punch cards so the impetus was never there to introduce a new technology system. It was other 'less successful departments' that were considered as higher priorities. However, it is now recognized that the new technology is required because there is a real danger of being left behind their competitors and having restricted access to information. It is also interesting that the manual system of punch cards was largely only effective as a result of the number of staff operating it, which points to a cost saving rationale in terms of the labour efficiency. Also important was consideration of the waste of the human resource since a potentially skilled labour force was undertaking jobs involving tasks significantly below their ability.

While the aim at Achilles seems to have involved improving information flow many of the justifications of managers refer to improving customer service, others refer to reducing numbers employed, to keep costs down. In Apollo, a company with similar aims as Achilles, justifications in terms of controlling staff are common.

The personnel manager at Apollo explained that the new technology developments are driven by requirements from the parent company to produce higher profits, new systems are one way of reducing costs and producing more. The financial accountant also reflected this view, and in keeping with this starkly capital orientated attitude the accountant went on to suggest that while there had been reductions in staff in the department, new technology has greatly enhanced managerial control.

Consistent with Storey's (1987) findings this has in turn meant efficiency gains have been realized. The invoice supervisor agreed with this explaining that it was possible to tell who entered information into the system and when it was entered so mistakes can be traced to the guilty party. She was, however, quick to point out that the system was used constructively and not punitively. Nonetheless, the possibility remains of using the system for surveillance and control to test employee performance. In the payroll area, staff had been reduced and the service offered had been significantly enhanced since a computer system had been introduced.

The systems manager justified new technology on the basis of its capability to reduce the number of staff. The operations manager of the manufacturing services division was following a similar line in his description of the rationales for the production control system: '....what we've been trying to do is by using the system and by better management, is to get our costs down and one of the prime ways of getting our costs down is you improve the method of manufacture but also you don't employ as many people....' This had meant a loss of clerical jobs in the order of eleven per cent. The loss of jobs was not entirely attributed to the stock control system (MANMAN) but the link was acknowledged.

Control over the conduct of staff has increased. The 'chaser' is still doing the same job but now he/she has the computer to assist. But the control aspect is also that the work is more disciplined, and this means that instead of the chaser getting an order through by using inducements ('a fag for the foreman') the production control looks after the sequence of jobs by due date, the chaser must first convince production control that an order should change priority in the listing. Management, of course, have direct access to the system and in this way can supervise the operation.

Summary - rationales

In each metalworking case the introduction of new technology is accompanied by job reform. While at Zeus new technology is justified on the basis of job changes, Artemis tend to avoid any direct association. However, other changes such as faster turn around times enabled through 'unmanned running' and 'multi-manning' at Artemis are seen as rationales for new technology. These changes are directly linked to work reorganization. Artemis management suggested that workforce flexibility was an essential precursor for new technology change while at Zeus maintenance craftsmen were retrained to become more flexible because new technology caused a blurring between electrical and mechanical trades. Zeus also rationalized new technology as a means of reducing the number of workers in the business and therefore reducing costs. At Artemis, cost saving was no less important but other rationales of quality and reducing lead times were mentioned, and these may have no bearing on saving money.

The newspapers show some signs of similarity with the engineering firms in their justifications for employing new technology to attain greater productivity and to remain viable and competitive. In terms of working practices both newspapers are seeking greater flexibility which inevitably means that the newspapers are looking to free themselves of trade union controls. An important part of the changes to a direct entry system is the change in the nature of the work which now requires little skill, without removing superfluous and duplicated functions the companies would find little advantage in having the additional technology. Here the rationales may be towards greater productivity and quality, but to arrive at that position traditional trade union controls must be removed.

The clerical studies differ from the other companies studied in their use of new technology. The tendency to use new systems as a tool to offset labour and skill shortages or prevent a burgeoning of staff, is common. Alongside this the improvement of service and greater capabilities have introduced new faculties that even extra staff would not have been able to provide. Nevertheless, the possibilities for using new technology for purposes of control cannot be ruled out, since the new technologies offer fewer opportunities for staff to discuss problems at work on a personal basis. In many cases systems can also identify users and gauge user performance. Many of the rationales put forward, deviate from the aims suggested in the previous section, suggesting that managers redefine or formulate their own aims and objectives independently of corporate decisions.

The previous sections have asked how and why questions. How is technology introduced and why is it introduced? The third question linked to these are: who is involved in its introduction? Influence through involvement is capable of informing and directing the very nature of change.

Involvement in new technology development

In the change to new technology different firms will consult different members of staff. Those who could be involved are trade unions, senior managers, line managers, supervisors, workers themselves, parent companies, and employers' associations. These actors may have important influences on the outcome of new technology introduction. Failure to consult with key groups, or unwelcome intervention of these groups may affect the way jobs are defined.

This section considers the involvement of various groups of actors in new technology programmes, the one exception is trade union involvement. Because of the potential importance of trade unions a separate section in chapter seven is devoted to their participation.

Metalworking studies

The section on strategy at the beginning of this chapter demonstrates that it is technology that receives the closest attention when planning for change. Jones (1988) has pointed out that discussions for new technological change very rarely include workers. Both Artemis and Zeus are no exceptions to this, and tend to confine their discussion about technology and jobs to higher levels in the organizational hierarchy.

In Zeus, decisions about the way jobs are designed are taken by the relevant production manager, operations manager, and senior unit supervisor. But a decision would first be discussed with all the machinery managers, although the gradings are controlled by the industrial relations department. The new technology changes, that were recently introduced, were based on a number of options which actually involved different combinations of employee grades, so there is a degree of interaction with the industrial relations department.

Decisions for machine purchase tend to be handled by a technical team. The production manager and operations manager would be involved in the decision making, but the capabilities of the machinery, and the decision to buy, rests finally with technical personnel. Some operators may visit a manufacturer to see the machine being made and for briefing, and then be closely involved in 'proving out' until they take over completely. However, many decisions involve major investment considerations about the whole plant, and concern the allocation of funds to one area or another. For example, whether to put money into assembly or machining. Such decisions are centrally based and lie with senior management. While the purchase itself is being considered the production management communicate with the technical department, and consider the repercussions of the machinery for job design and grading. Here again the instrumentalism of Zeus's managers is illustrated, the main concern was that machines could be implemented '....without causing us to introduce grade B plus or something silly because of the complexity of the machine.....' (Production Operations Manager, Zeus).

However, an interview with the AEU convener suggested an alternative scenario: '....a lot of times it seems to be a suck it and see situation. Sometimes they find that more people are required to operate it.' Discussions with management did take place and they had advised that there were a set of machines planned for introduction, but there was uncertainty about how many people this would require. Despite this, at least one production manager at Zeus was willing to argue the positive aspects of consultation with workers prior to implementation of new machinery. There is an increasing tendency, he suggested, to discuss new technology locally '....whereby there is a lot of involvement from the shopfloor to make sure we get their ideas, you've really got to take the people along with you at the outset.'

Perhaps his concern was due in part to the fact that he had recently been promoted from foreman, and consequently was more aware of what was happening on the shopfloor. Nevertheless, he did point out that people are sent to the machinery manufacturers for training and they later prove out the machines. In this way it can be claimed that they are involved at the planning stage. Another part of the consultation process was carried out in improvement teams. He claimed that the reason for asking people lies in smoother implementation and less modification later on.

At Zeus main decision making involves senior management, technical decisions are taken by technical staff in the technical department. The industrial relations department is brought in where there are grading issues to be resolved. Generally, two views were expressed by general managers on the involvement of shopfloor workers in discussions for new technology. One group wanted to consider job design issues because, with the addition of new tasks, they felt that higher grades might be forced on them as a trade off for workers fulfilling those tasks. The alternative view felt that workers' views were important in their own right because of their knowledge of the work.

At Artemis, the personnel manager, machine shop superintendent, manufacturing manager, and industrial engineering manager discussed the changes in technology in the organization. The presence of the personnel manager points to some awareness of the effect on jobs.

The production director takes the evidence from this group, devises a rationale, and passes it to the board of directors for a final decision. Managers claimed that there were also workforce discussions, held within quality circles or through informal contact, about how the various flexibility changes were to be implemented, but these were limited and not always successful. Discussion with workers seemed to have little to do with dialogue and with receiving workers' views, but had more to do with managers advising the workers what was to happen.

Otherwise, managers asked the supervisory staff for their impression of the general workforce's feeling. Nevertheless, proposals had provoked a strike over the removal of individual bonus payments, and there was also concern about what would happen to inspectors with the introduction of self inspection. Members of the workforce interviewed expressed the view that, both the industrial action, and expression of concern, had seemed to impress the management. The evidence suggests that these were issues that could have been resolved without strike action, had consultation been more extensive. But the fact that strike action had been used perhaps highlighted how remote effective consultation was.

Artemis shows evidence of some contact with workers but the form this took was more of managers advising workers of what to expect. Industrial action demonstrates the inadequacy of this. Artemis is a smaller company having fewer layers of managers so here responsibilities are spread more evenly in the organization. For example, in Artemis there is no technical department so technical feasibility is considered by the industrial engineering department.

At the newspapers, the choice of technology and job designs were even more firmly in the hands of management.

Newspaper studies

Managers at the newspapers were unconcerned with consulting the production workforce, although some attention was paid to the views of editorial staff. While at Hermes, parent company intervention played an important role in the change process.

Job design was not an issue of great discussion at Mercury when the company went to cold composition from hot metal, since managers felt that little had changed. Little retraining was required to take compositors from linotype style keyboards to QWERTY keyboards, and some retraining took place in the pre-press area and in the change from making papier mache moulds to bromide paste up (see chapter four). The staff remained the same and doing the same functions, albeit with different equipment. The changes to direct entry are less involved but mean more changes in jobs and more job losses in the composing area. The main decisions were taken by the production manager, senior editorial staff, the advertising manager and the managing director. The details that were discussed involved the level of training, the number of staff required, and the system that was to be used. The editor decided on the job design for the journalists, and the assistant editor was involved in discussions at times when the jobs involved the direct entry system.

The purchase of the system at Mercury was discussed mainly by the advertising and editorial department heads, and in the process of making the decision, staff were asked what they wanted and needed, for example, how many terminals were required. The asking paid dividends, according to the assistant editor. The system became closely tailored to the needs of the staff and the newspaper. With the exception of the purchase of a new press the parent company did not intervene at all, and a system was chosen that had full expansion capability. However, on the production side of the business any workforce discussion that took place was in the form of negotiation, and based mainly on issues of the number of redundancies, redeployment and levels of pay, issues discussed further in the next chapter. Job content and work organization were non negotiable issues, and in any case were considered as set by the technology, as discussed in earlier sections of this chapter.

No discussion took place outside trade union agreements between the NUJ and NGA (trade union involvement is discussed in chapter seven). However, the editorial management seemed much more prepared to discuss prospective changes with the editorial staff, than the production management was with production workers. Hermes experienced a similar approach, but the process was complicated by the initial presence of the Newspaper Society and subsequent intervention of the parent company.

During the initial interviews at Hermes, the production manager painted a picture of a newspaper very much led by the managing director, who in turn relied heavily on the production manager. The production manager had been responsible for the changes to photocomposition, at the managing directors request, and for reasons of his experience. He in turn delegated the training to the overseer in the composing room but undertook aspects of training himself. When it came to buying a new printing press, Westminster Press intervened directly, introducing an 'adviser' from their technical services division. But once the decision had been made, the production manager again stepped in to handle the training and to reorganize the site.

The relationship between the managing director and the production manager is clearly important and interesting for this case study. Westminster Press, the production manager informed me, would never get embroiled in issues involving staffing or negotiation. These issues were generally the domain of the managing director. Hermes, like Mercury is a member of the Newspaper Society (NS) but unlike Mercury, Hermes used the NS in negotiations. This was not discovered until I spoke to the NGA FOC. Clearly the production manager was not prepared to release the information. On a previous occasion he had told me that the Newspaper Society would only be used '....where we'd exhausted procedures locally. It would be getting pretty serious at that stage.'

Yet procedures had hardly begun. Both newspapers went through 'talks about talks'[14] because neither wanted joint negotiation with the NUJ and NGA, fearing that this could have crippling results for the newspaper if there was a failure to agree. Despite this, the newspapers eventually agreed to accept joint negotiation. As soon as this initial breakthrough had been made at Hermes, the Newspaper Society appeared at the negotiating table in stark contradiction to the declared policy. The climate of negotiation soured considerably. Westminster Press, the parent company, stepped in. The existing managing director resigned, and a newly appointed managing director took his place. The negotiations were suspended for one month, while the new head familiarized himself with the affairs of the newspaper and state of negotiations. Negotiations later resumed, but without the Newspaper Society.

Simultaneously, the production manager was relieved of his responsibility of leading the direct entry changes, which were handed to the editorial side of the business and the deputy editor. The production manager claimed he had new responsibilities to concentrate on, but it was clear to many people involved in the negotiation that he, with the managing director, had not handled negotiations well. The NUJ FOC reported that the management of Hermes generally had a very casual style, but always tended to emphasize the independence of the operation. He had little doubt that Westminster Press had become directly involved, although the extent of the involvement was unknown. It was clear from these events that Westminster Press would get involved in the running of newspapers if they felt there was reason to.

The ultimate approach at Hermes was close to the experience at Mercury. However, the approach of Hermes was initially quite different. The idea of forcing change, and avoiding consultation, brought parent company intervention and the replacement of the managing director.

Office studies

In Achilles different groups of managers relied on each other to deal with staff consultation. As a result, the real responsibility for staff consultation was confused, and managers seemed to be satisfied with an approach in which they were seen to be consulting, but in reality little was actually taking place. Managers at Apollo failed to see, or to acknowledge, the relevance of involving workers, or questioning the extent to which they could make valid suggestions.

While the major decisions of what system to buy, and where to place it in a list of priorities, fell to the board of directors after cost justification, Achilles tended to emphasize the involvement of staff interviewed by systems analysts, who were conducting specific projects in the earlier stages of decision making. The systems manager also pointed out that when the analysts' report was complete, managers would normally pull their staff together and discuss the issues. He also referred to the advantages the company might gain from employee innovation in the operation of a system, and the more negative, but nonetheless essential side of discovering pitfalls. However, in several cases managers claimed that it was the responsibility of systems personnel to discuss these issues with the staff. The confusion over whose responsibility it was, inevitably meant that no one talked to the staff.

In the distribution department new jobs were discussed by the key managers in the business. These included the warehouse manager, the distribution and customer manager, and representatives from the management information systems department (MIS), and these discussions produced broad specifications. The discussions then filtered down to lower management levels, cascade style, and involved issues of the forecasted change, the jobs, and the equipment. The managers then began to organize their staff to meet the change. Staff consultation was in the form of discussing different plans at different stages of change, often breaking up into project user groups to do this. However, once again the view was adopted that for the most part the system dictated what the job would be, and as a result there was clearly little room for designing work with interest and involvement: 'The jobs almost, in a sense, make themselves, being led by the equipment that you are buying and the computer systems that you are using....' (Distribution and Customer Manager, Achilles).

Any involving and interesting work was a result of a managers' particular interpretation of how a job should be designed to meet the system requirements. Managers would package jobs in accordance with their own particular set of ideas and logic, including technological determinism. They failed to recognize that there were a wide range of alternative possibilities available for the efficient organization of work.

The credit manager at Achilles explained that he was involved in the choice of system for the accounts department, with the invoice supervisor, two technical people from the (MIS), and the financial accountant. This group met as a project team to discuss the viability of computer packages and also visited companies that produce the product. When discussions have narrowed the choice down to two packages the staff are informed of the likely outcome by cascade briefing. Ultimately the financial director will discuss the details of the packages with the financial manager and a recommendation will be made to the board. The financial manager made it clear that staff were not considered or consulted in the course of deciding which package to buy (see also Wainwright and Francis, 1984).

In Achilles, systems staff claimed that managers were consulting workers and advising their views. In some cases managers believed that it was systems personnel who were talking to potential staff users. The reality is inevitably that very little staff involvement materialized. Decisions were made at senior level involving managers and system staff. In most cases, line managers and staff only knew of the changes was when the information filtered down to them. In addition to this, technological determinism was common in excusing any lack of consultation.

Apollo, like many of the companies studied, had a structured system of decision making for the purchase of equipment. Systems were discussed in committee, and the final decision was taken at local board level. In the accounts department of Apollo, the accountants decided that there was a need to improve the systems that were in use, and two managers from the accounts department, and one from MIS decided on the new system to be used. As in Achilles, it was deemed inappropriate to ask the staff about their jobs because the management team felt they already knew the jobs intimately: '....when I talk to my staff it's not a case of me learning what they do, I already know....' (Financial Accountant, Apollo).

This failure to acknowledge the fact that there are aspects to individuals' jobs that only the workers themselves aware of, did not reflect in the general approach to consultation with staff. The same person then launched into rhetoric about team working and the importance of involving staff in changes to new technology. He went on to deliberate on the rights and wrongs of staff involvement:

> ....if you impose a system that doesn't work you've got problems, if you impose a system you will normally have opposition anyway. If it is a system that they themselves have evolved and been party to the decision making process you don't get opposition and it's a very effective way of harnessing your resources (Financial Accountant, Apollo).

However, he claimed it was necessary to set general rules for working which should include no pay increases and no promotions. Clearly, despite the rhetoric, there was likely to be little that the staff were able to influence or gain from the changes, unless their wishes coincided with those of the managers, and the role the managers had assigned to the technology. Nevertheless, most of the staff interviewed suggested that they had been consulted and felt they had some opportunity to comment. Much of this discussion took place with the systems accountant who was appointed to look into computerizing different areas of the accounts department at Apollo.

The operations manager in the MSG tended to highlight one of the practical problems of asking the users what they required. He claimed that people simply have no idea what they do not have, so when they are asked, they have difficulty in articulating their requirements, apart from the obvious things that they are already used to. However, this seemed to be a good excuse for not having to think too much about user requirements or job designs. The obvious solution to such a problem is to advise the workforce of what is on offer and allow them to make decisions on that basis. Clearly this lack of imagination has led to the entrenching of exclusive managerial control of work organization. For the managers, the lack of staff involvement did not create great concern since the changes were not seen as highly significant. They were again viewed as merely the use of different tools to do the same jobs.

At Apollo, different departments reflect different approaches. One approach considers worker knowledge to be irrelevant. Any involvement of workers was to avoid further opposition. Another approach was simply to deny the usefulness of consulting workers on the basis of lack of knowledge of the new systems. Again a common feature in this company was to see new technology as determining job designs and therefore to deny the need for any consultation.

Summary - involvement

The two metalworking companies demonstrate similar approaches to decision making with new technology systems. Inevitably, capital equipment investment involves heavy expenditure and warrants careful consideration in terms of cost accounting, and balancing the cost against the benefits that are likely to be received. The level at which this is conducted is for both companies closed to anybody below senior management level.

Job design is too often a retrospective consideration. Thought only being given to this when the technology had been chosen. Even when staff were involved in discussions on issues of job organization, little of this seems to have been taken into account.

Hermes and Mercury newspapers have many similarities. New technology decisions and job changes were discussed with the editorial staff, but the destiny of the production staff was already decided and the systems being introduced were to execute a change in work organization which was strongly opposed by those in production. Where Hermes differs considerably is in the involvement of the Newspaper Society in negotiations, and subsequently the intervention of Westminster Press. The former reflects a heavy handed management style, the latter an opposition to such a style. The remaining manufacturing companies promoted an image of far more worker involvement, but in reality this was limited.

Achilles seemed to recognize the importance of being seen to consult staff and users, but their real commitment to this was incomplete and this resulted in confusion about who was really speaking to staff. In both the distribution department, and the accounts departments, the important discussions were between managers, and for the most part, the staff were only advised rather than consulted about changes. In Apollo there was opposition to the inclusion of staff in one department. In another the importance of staff consultation was recognized, but not realized because managers doubted the ability of the staff to make a contribution due to their lack of knowledge. However, the manager of this department had overlooked the fact that if the latter group was better informed by managers, the validity of a contribution would not have been in doubt.

Summary and conclusions

Companies introducing new technology in pursuit of specific aims and objectives are likely to plan organizational changes. However, strategy may be incomplete (Rose and Jones, 1985; Child, 1972). For example, consideration of technical requirements may take place to the exclusion of work organization. However, corporate aims require departmental cooperation if they are to be realized. In this process, line managers define their own justifications (rationales) for new technology usage. These may be responses to the poor planning of senior managers or may diverge from the overall aim. Therefore, groups of departmental rationales emerge, some of these are related to the corporate aims, others are quite different. However, managerial perceptions of achieving aims will be likely to effect the way that jobs are designed.

Finally, also of key importance in the process of change is the range of influences that affect decisions. These include decisions about setting objectives during the process of implementation. The section on involvement looked at the procedure for choosing new technology and organizing work with particular reference to the involvement of the workers and users in the process. This is an important question, not only for the workers, but also for corporate interests. Worker knowledge, can help in job definition and in the choice of new systems and machinery because they are aware of the component parts of the jobs they perform.

Confirming the hypothesis, the office and newspaper studies illustrate a tendency to excuse the absence of job planning. They achieved this by assuming the new technology, which had been considered carefully, would by itself shape and dictate the way work should be organized (an issue further discussed in chapters one and two). Managers at Artemis, the manufacturer of precision hydraulics, felt there was no need to consider job changes in connection with new technology, since workers were already operating new technology, so only a few incremental changes would be necessary. At Zeus (the diesel engine manufacturer), planning took the form of preparing job descriptions so that workers would not have legitimate claims for pay increases.

The newspapers adopted similar approaches to their counterparts elsewhere in the industry. But often, the peculiarities of the particular companies involved were not accounted for, which prompted some workers to suggest that changes were still under consideration, implying lack of advanced planning. Similarly, at Apollo (the aerospace company), some planning for job changes proved insufficient and resulted in later incremental changes. In one area, Achilles (the shoe company) seemed to be encouraging a flexibility in some of its workers that would relieve managers from carefully defining jobs. Nevertheless, both of the office cases planned new technology systems very carefully.

In accordance with the hypothesis, planning in each of the cases was limited. In most, planning for technology was a significant concern, while planning for jobs was secondary, incomplete, or non existent. Where planning does exist, it is often to avoid conceding pay increases or to offset opposition. There is no evidence of a sociotechnical approach in which job definition has more humanistic goals.

While the rationales offered for new technology introduction are varied, and often deviate from the more general aims in the company, there is evidence that some managers see motives of control over workers as a justification for its use. Possibilities of the use of new technology for surveillance and control are evident at Artemis, the newspaper companies, and in Apollo. Elsewhere other managers would place quite different emphasis on the introduction of new technology in accordance with their own concerns. The hypothesis on rationales can only partially be accepted, since there is evidence of new technology used in accordance with motives of control, although this by no means predominated.

In no company was the choice of new technology considered to be an area that staff should be consulted about (see Cressey, 1990; Cressey *et. al.*, 1988; chapter seven on trade union involvement). Often involvement relating to issues affecting jobs was conceded only to avoid conflict. At Zeus and particularly Hermes, parent company intervention directly influenced the outcome. Also direct contact with workers here, and at Artemis, did little more than advise people of what was happening in the organizations. At Hermes, where the Newspaper Society was also involved, the parent company prompted changes in senior management and the direction of the negotiation procedure. In the office studies, involving workers either seemed irrelevant, or managers doubted the practical contribution that workers could make. Levels of discussion, as the hypothesis suggests, were generally confined to senior management.

The evidence generally supports the hypothesis although there is evidence of control over workers being included in manager's rationales. Rationales, nonetheless are varied and reflect lack of overall planning. At least where work organization is concerned, strategic plans for change are a rare consideration and discussions that take place may involve a number of parties but in most cases workers are excluded from making constructive contributions.

The next chapter considers the involvement of trade unions as a key and potentially influential group in procedures involving the introduction of new technology.

Notes

1. It is also interesting that the manager should use a figure of 500 redundancies so freely but in Zeus the sad truth is that large numbers of people left the firm when it found itself in crisis. Hence we have to be careful about attributing this sort of volume of job loss to technology changes, for the most part it was organizational change, but judging by the technological progress Zeus had made and

continued to make, technology was a key feature of that organizational change.

2. Another example emerged of the effectiveness of this approach when a new robot operated cylinder head line was introduced to the factory. The line was staffed with a 'hand picked team' of operators who went to see the machine being made in Bristol where they received some special training. The upshot was that 'the team' requested a grade increase but by virtue of the company's forward planning and anticipation, the new tasks had already been included in the job specification. In any case, the management argued, the jobs were now more simple than before; no grade increase was conceded. This manager had (perhaps not surprisingly) established as one objective for the year, the capability, as a department, of absorbing new technology without the workforce refusing to run it and without paying more money.

3. See Chapter Two and particularly the section on the sociotechnical approach and discussion of technological determinism.

4. Similar arrangements were made at Hermes, although with smaller financial resources Hermes had to fall back on the possibility of compulsory redundancy. In the event this was not necessary.

5. See chapter two which identifies the possibility of conflicts within and between trade unions; an issue taken up in chapter seven.

6. Interviews took place before the system was implemented.

7. There were VDUs liberally placed in the offices and on the shopfloor. One purpose these were put to, was the location of the components that made up particular orders. Large customers had VDUs linked to the Zeus' system for the purpose of seeing the progress of their own orders.

8. See chapter seven where the demarcation issue is discussed fully.

9. The denial of new technology as a cause of change seems to be quite common and possibly derives from trade union agreements that insist new technology should not cause job loss. This provides little problem for determined managers who simply call it 'organizational change'. Preliminary enquiries made of British Telecom, for example, elicited precisely this response.

10. See also Cockburn (1983) on the Croydon Advertiser.

11. In both Hermes' and Mercury's regions, independent 'free sheets' had been established for some time, taking advantage of the relative cheapness of new technology systems. This threatened advertising revenue as indeed did other forms of advertising media such as local radio. However, the NGA argued that these claims were not so convincing since ownership of many of the 'free sheets' and other forms of

media was in the hands of the newspaper companies themselves.

12. See Cockburn (1983) on the significance of skill shifts in the industry and Rainbird (1988) on the redefinition of skill.

13. Although this is a highly sexist statement on the part of the overseer, regretfully it probably is not too far from the truth. Mercury had begun to take on casual staff or short term contract staff. They would have few rights under the law and there are few unions equipped to recruit this kind of workforce even though both SOGAT and the NGA would take them given the opportunity.

14. 'Talks about talks' refers to pre-negotiation discussions about the way talks were to be conducted. For example, a central issue was the question whether negotiations were to be conducted jointly (NUJ and NGA) or separately.

7 Trade union involvement and demarcation conflicts

Hypotheses

1 - Trade unions primarily associate new technology with quantitative issues such as job loss, severance pay and new pay levels. Lack of knowledge and resources lead to inability for making a case for trade union inclusion in discussions of the qualitative effects (work organization, job design) of new technology.

2 - As new technology is introduced, occupational mergers take place which may mean the breaching of traditional demarcations. Where this happens conflicts will develop within and between trade unions.

Introduction

Chapter six has considered various forms and levels of involvement but these excluded trade unions. This chapter deals with trade union issues of involvement and demarcation arising from the introduction of new technology.

If new technology is to be implemented successfully, and work organized appropriately, those making the decisions need to be fully aware of what the work actually involves. According to one theory of change management, this can be achieved by involving the workers and union organization in the discussion for change. This relies not only on the extent to which managers are willing to allow such discussions to take place, but also on the strength of a trade union's case for inclusion. Trade union effectiveness is criticized due to poor performance in this field (Wilkinson, 1983). But others have shown that trade unions are able to have real effects on managerial policy making in areas of work organization (Batstone *et. al.*, 1987). The trade union involvement hypothesis is based on the premise that trade union organization at local level, is neither sufficiently adept, nor resourced, to be involved. Where trade unions are involved, they fail to tackle issues such as job design and work organization, but tend to deal with more tangible and immediate issues such as loss of jobs and pay.

Evidence from the case studies indeed demonstrates that trade unions fail to make effective cases for inclusion in new technology discussions over work organization issues. However, the hypothesis can only be accepted reservedly, since it is clear that managerial attitudes towards trade unions play an equally important role in the process. Even where trade unions are well organized, management intransigence can prevent any constructive use of this potential resource. In some cases, managers accept the importance of talking to trade unions for political and practical purposes, but are unwilling to compromise through bargaining. Cressey (1990) has suggested that there are a range of levels of union involvement in new technology decisions. These extend from no involvement, through to notification, consultation, negotiation and joint decision making. The least involvement is at the planning stage, which is in line with findings in chapter six. Participation increases as there is movement forward into the implementation stage.

The most important distinction made here is between consultation and negotiation, the former refers to discussion between managers and unions, the latter refers to bargaining.

The second hypothesis, considers the consequences of job change, in the form of occupational mergers that breach traditional union demarcations. When new technology is introduced there are changes to jobs. In the case study companies these changes have tended to follow the addition of new skills. Whether this is a vertical or horizontal addition, the new skills are often acquired by merging jobs. Occupational mergers inevitably breach traditional demarcations, and where this occurs conflicts are likely between workers affected by the change, and between trade unions who represent the workers. Further problems arise when trade unions represent one or more occupational levels that are merged in this process, which potentially creates internal conflicts.

There is indeed evidence that conflicts are developing between some trade unions and workers in the case studies, although this is not as strong as expected, and seems to be concentrated more in certain sectors. Hence the hypothesis is accepted only partially, since the evidence is scattered. It is, nevertheless, essential for unions to recognize the potential importance of the development of such conflicts and develop responses accordingly.

Trade union involvement, job design and new technology

One management point of view reflects the view that unions are necessarily opposed to the objectives of managers. However, this need not be the case. There is considerable evidence to suggest that trade unions often see new technology as potentially beneficial (Rainbird, 1988; see also Davies, 1984). Jones (1988 p472) has claimed '....trade unions have given considerable official support to the introduction of the technology of flexible automation....' Batstone *et. al.* (1987) amongst others, have documented the case of a trade union who identified new technology with both business success and gains for their members.

Metalworking studies

While managers at Zeus, the diesel engine company, recognized a need to consult with trade unions, this was undertaken only to avoid confrontation later. Managerial preferences were towards a style of consultation that went direct to the employees. Industrial relations at Artemis seemed largely based on personalities. Managers achieved little with a new senior shop steward, although considerable progress was made with the former shop steward. Decisions were often made in advance of consultation, and a strike revealed how ineffective consultation really was.

Managerial attitudes

The management at Zeus seemed to be operating in a way that would prevent any concessions to the union, even where such concessions would have possible beneficial effects on the management of the business. The production operations manager suggested that often decisions were made first, and only then discussed with workers. He viewed trade union involvement negatively, and the only purpose he could see for this brand of 'consultation' was to ensure that there was no trade union dispute over lack of consultation and advice. A statement confirmed this position:

....another technique that we use....is the controlled leak. What it really is, is telling somebody about something as far in advance as possible, it's so far in advance that if they object they don't take action about it, all your telling them is something is going to happen (Production Operations Manager, Zeus).

The same manager conceded that a degree of honesty was required when production engineers were on the shopfloor making enquiries and taking measurements, because this was so tangible and would clearly raise questions.

The personnel manager at Zeus claimed that the company was committed to trade union consultation, but what actually happened did not depend on agreement arising from that discussion. Generally, Zeus managers seemed determined to avoid concessions to trade unions, and wished to confine any talks to 'consultation' with no commitments to respond to suggestions.

Labour relations at Artemis, the precision hydraulics company, did appear to be satisfactory but the history of the company revealed a strike and a management lockout over the proposals to change the bonus scheme. Since this time the senior levels of management had changed. But, nevertheless, the new manufacturing manager displayed an uncompromising style. He did claim however, that he saw the benefits of union consultation and referred to the extensive negotiation that took place when the grading system was changed. The major reorganization that had taken place in work organization and new technology was 'thrashed out amicably' with the previous senior shop steward (he had since died), the personnel manager and the manufacturing manager.

From the way that the manufacturing manager and personnel manager spoke about the old senior shop steward, it was clear that there had been a good deal of trust between them. This assessment of personalities proved to be important, and contributed to favourable conduct of relations. There was little respect or trust for the new senior shop steward, who the manufacturing manager claimed was 'no good' because he had to refer to the branch office for advice and could never make a decision by himself.

Consultation procedure

Zeus had developed a method of using team brief and improvement teams (quality circles) and going direct to the workforce rather than via the trade union. Joint consultation did take place between the unions and managers but it would only occur alongside direct communication with the workforce. 'Team brief' is a cascade style briefing, and it was pointed out that using this system was not possible when the union stewards would receive the information before the foremen. But there was even a strategic side to this technique. Telling people individually, or in small groups, presents far less likelihood of problems, than informing large groups which would be the scenario at a union meeting. There was another argument for a reassertion of managerial prerogative: 'It gets you back to the foreman managing the patch again, whereas before the font of all knowledge was the union....the man goes to the foreman if he wants to know something and not to the shop stewards' (Production operations manager, Zeus).

Improvement teams (quality circles) were strongly opposed by the union, team briefs were tolerated. The union claimed that through the non cooperation of workers, many improvement teams have been disbanded. The trade union felt aggrieved by the fact that the management wish to consult directly with the staff and bypass the traditional channels of communication. One of the shop stewards saw the teams as a means of doing away with shop stewards, and ultimately the union, and replacing it with a form of 'enterprise unionism.'[1] Another steward explained that team briefing had other advantages, claiming '....the lads look forward to team brief because it's half an hour off the job.'

The operators and union officials complained that new technology systems, generally, just appear, and the workers concerned have to try learn the functions and cope with them. There was evidence, however, of a positive move by the company to involve the trade union at an early stage in a very broad range of issues. However, this attempt at bringing the union into more of a decision making capacity, was only to be permitted, if the unions agreed to the condition, that some of the information they were given should not be passed onto the workforce. This was because there were likely to be sensitive issues involved in discussions, that would be valuable to the company's competitors.

On moral grounds the union felt it could not be party to information that would then have to be withheld from their members, and they declined the managements' invitation. In doing so they probably missed an important opportunity of being able to influence management at an early stage. Batstone *et. al.* (1987) refer to a similar case, where the trade union took full advantage of a similar offer from management which enabled anticipation of problems, and in one case had revealed a new technology programme that had not been discussed with the workforce. However, at Zeus the experienced union representatives were suspicious, and doubted that managers were genuinely promoting a high trust relationship (Fox, 1974). It was likely that managers were attempting to capture a high trust response, benefiting from union knowledge and innovation, but in what was an inherently low trust situation (Sabel, 1982).

The Zeus Era factory seemed to be governed with a more congenial style. Due to its size, everybody seemed to know everybody else and reports of effective consultation were more common than at the main site. Although these were still on the low side, and conspicuously excluded job design issues. They dealt more with working conditions and factory layout. Management was seen by the workforce as approachable, and relations were reported as reasonably good by the trade union. This apparent harmony needs to be seen in context however, the management at the site report a very passive union who tend to take the 'easy line' with situations.

Zeus' management was making considerable use of methods of direct worker notification and consultation that bypassed the

Zeus' management was making considerable use of methods of direct worker notification and consultation that bypassed the trade unions. This was clearly intentional and meant the company could hear workers views and minimize the risk of opposition, while at the same time not committing themselves to responding to workers' contributions. Artemis had also begun to develop direct methods of consultation.

Consultation with the workforce at Artemis seemed to operate at both the formal and informal level. The production director spoke of going onto the shopfloor and 'grabbing some workers' to discuss issues with them, a direct consultation that the senior shop steward did not object to. Where more formal, effective consultation had taken place was in the use of the knowledge of the superintendent in the machine shop. The manufacturing manager, and the superintendent himself, reported long discussions about the issue of general work organization through to the suitability of individual workers for particular jobs. However, in the new build assembly area the superintendent complained he had not been consulted at all. All the discussion had been carried out between senior managers and his superior.

The personnel manager explained that there had been extensive consultation with the shop stewards prior to the main organizational and technical change:

> From my point of view....the worst thing anyone can do is to just bring in change and new machinery without telling anybody, whereas the art of negotiation is to get the other side to see your point of view. Management has to take the workforce along with them in the acceptance of change, therefore prior to 1983 we held a series of consultative meetings with the trade union representatives....in order that everyone was clear what was actually happening.

The format of the meetings, was a review of what had already taken place, and consideration of what was to occur in the future. Discussions were carried out in what the personnel manager described as an informal atmosphere, and the importance of involvement of workers on the shopfloor was not overlooked. But the company, unlike Zeus, opted to use the traditional method of using the shop stewards to represent the workforce. The main aim of any consultation was to reach individuals, but shop stewards were used as intermediaries to receive workers' ideas, present these to management, and report back. The personnel manager, who was clearly very enthusiastic about what had been achieved, claimed that: 'everything we did, we really did it in concert with the trade union, and why not?'

This statement was patently untrue, the trade union had been extensively involved in negotiations, but the manufacturing manager pointed out, that the decision to make staff redundant was solely a management arrangement. What had been discussed was redundancy levels, but these negotiations took place long after the initial decisions were made. One of the senior CNC operators also pointed out that the workforce was informed about the main job changes only 'a couple of months beforehand', and there had been no consultation at all about the new machinery being introduced. For a machine just to appear was a common occurrence, and created considerable anxiety amongst the workforce who were uncertain about its purpose.

Trade union approaches

According to the production operations manager at Zeus, the only times that the union did get involved in discussions over new technology introduction, was when they wanted to increase the gradings (see Wilkinson, 1983). However, given the fact that they were given little opportunity for involvement this seemed to be hardly a fair statement. There was clearly a dislike of the trade union organization, and with the tape recorder turned off, lest I played it back to the shop convener, I was told of the unacceptable intransigent attitude that management had to put up with. As an example the production operations manager reported that even health screening to anticipate industrial accidents was rejected. Obviously, from a union point of view, this was on the basis that the management may decide that certain members of staff should be removed because of their poor health. Based on the general conduct of relations, it did not seem an unreasonable assumption.

In the case of the maintenance group, the union accepted the need to retrain readily and did not attempt to define the jobs as such, although there were moves amongst the electricians to retain aspects of electrical work as exclusively theirs. As noted in chapter five, the union did insist on a 'no embarrassment clause' to protect anybody who felt humiliated by not being able to aspire to the higher levels of skill required in the multi-skilling arrangement.

The works convener claimed that more involvement took place than the personnel manager was willing to admit to, suggesting most discussions centred on 'manning levels and how it is manned.' This turned out to be an allocation issue, the union wanting the worker with the longest service to get the best job. The management at the Era factory mused that allocation based on length of service criteria would result in the toilet cleaner taking the managing director's post.

Furthermore, in the case of redundancy, the unions' view was that selection should be made on the traditional basis of 'first in last out'. This criteria did not meet with managers' approval since they were determined to interview staff for any new positions, allocating points according to the 'quality' of the candidate. This same system was used (although with a more negative slant) when not enough volunteers were available in redundancy programmes. In the case of new technology introduction the convener explained that it was always considered that new equipment warranted an upgrade, but the case was often difficult when much of the new technology made the work simpler.

One trade union initiative at Zeus was an attempt to secure a new technology agreement (NTA) with the company. The company responded by asking for union cooperation to use temporary labour in peaks of business. The union refused to agree to any temporary or contract labour, so the agreement never materialized. The shop steward interviewed at the Era site argued that the more recent forms of new technology should not have been operated until the company had agreed to the unions' terms for the agreement. It is worth noting however, that quite apart from having the benefit of hindsight, this steward spoke from a safe conventional machining area, and had never been personally faced with such decisions on new technology. Other studies have shown that such agreements are rare in manufacturing industry (Williams and Steward, 1985). Where they are established there is often much discussion over union demands and, as was the case at Zeus, counter demands are often made by the company in response (Batstone *et. al.*, 1987).

The basis of the new technology agreement, from a union point of view, would have been to protect the established worker from being downgraded by a technological innovation that removed the workers function, and to protect jobs. To facilitate this it would also establish a policy on discussions with the union prior to any implementation decision. Like so many of the measures the union was most vocal about, there was no mention of work organization but a preoccupation with grading and financial rewards. From a management point of view this was typical, and was perceived as inflexibility and an inability to respond to the crisis that existed in Zeus.

The approach was indeed typical. In all the interviews with shop stewards, and the convener, it was the financial and grading issues that were prominent. All remembered the halcyon days of a wealthy company and powerful union, and arguments were generally based on what they had been used to which failed to take account of current circumstances. However, the trade union's position was not assisted by an intransigent management attitude, that would always make counter claims in response to a trade union request. At Artemis, the relations between unions and managers were generally less strained.

Overall, the consultation at Artemis seems to have been more positive than at Zeus, with the union expressing some interest in the redefinition of jobs, according to the production director's account. However, the only examples of this proved to be about the amount of the bonus scheme, and the enhanced terms the workers would receive for agreeing to work flexibly. Both of the superintendents, and the manufacturing manager, maintained that the union was concerned largely with financial issues, and were particularly worried about loss of membership and loss of hours. However, whilst the union was interested in better rewards and retaining their current benefits, it was revealed that the senior shop steward had put forward a case for a new technology agreement, that would lay some kind of plan for the implementation of new machinery. This was rejected by Artemis' managers because they felt that an agreement of this kind would be too restrictive.

Summary - metalworking studies

Zeus was prepared to speak to trade unions, only on the basis that consultation was not likely to be allowed to influence decisions. The trade union seemed preoccupied with financial issues and the management with limiting the power of the trade union. Suspicion on both sides scuppered any useful system of trade union consultation. Although at Artemis the senior shop steward was preoccupied with financial and grading issues, as at Zeus, requests for a new technology agreement had been made. This was rejected because managers feared the restrictions this may impose on them. However, informal forms of consultation had begun to emerge that bypassed the union. In both of the metal manufacturers, consultation about the new technology to be introduced was inadequate. The trade unions were preoccupied with maintaining levels of employment, a traditional hierarchy and, where appropriate, achieving pay increases in the face of new technology introductions. Little evidence was provided in either company, of the unions attempting to influence the work organization. At Zeus this has to be balanced against some effort of the management to include them in such discussions. Nevertheless, there was clearly no intention, in each case, of any discussion about the types of new technology to be employed, these were strictly managerial decisions.

At Artemis, despite a recognition of the importance of consultation the management failed to advise the workers of fundamental issues that affected them directly. Redundancies were decided on, and new machinery had been purchased, without any reference to the trade union or workers. In both companies, but particularly Zeus, trade union organization and management viewed each other in opposition. Other studies, notably Batstone *et. al.* (1987) have shown that it is possible to have a productive relationship in which both sides identify similar goals.

The hypothesis is confirmed here since trade unions seem mainly concerned with quantitative issues, but management approaches to union organizations had also been important in preventing effective dialogue.

The newspaper studies represent a similar scenario but the unions restricted their focus to financial, and maintenance of employment number, issues to an even more acute degree. There is also a stronger anti union feeling from management based on years of effective opposition from the NGA.

Newspaper studies

The newspapers display a much more direct approach than the metalworking studies. Here little time was wasted with the form of consultation experienced in the metalworking companies. Instead, the approach was for management to decide on the form of organization, and negotiate only its implementation.

Managerial attitudes

The production manager at Hermes (the smaller newspaper company) pointed out that, as at Zeus, listening to the unions' views, was not always 'agreeing or taking action.' In many instances, it was felt inappropriate to consult the union or workers, and in others, union suggestions and requests were not carried through. This, of course, is quite usual, but one proposal the production manager had rejected on cost grounds, was based on concern from compositors that there was not enough training available, when the photocomposition system prompted the change from linotype to QWERTY keyboards.[2] It seems unusual to reject such a request for training which would be likely to be beneficial to the company, especially since the production manager had also boasted a lot of proactive consultation, claiming his 'door was always open.'

Like many of the newspapers, the initial change to photocomposition maintained broadly the same distribution of jobs and workers, although job content had changed.[3] The direct entry changes mark the point at which these traditional demarcations were in jeopardy. By this stage the unions had a good idea of the kinds of new technology that were to be introduced, but in both newspapers management discussions about technology and work reorganization had excluded the trade unions. The negotiations that managers were prepared to have with trade unions, were against the background of technology and work design decisions having already been made. What was left for discussion were redundancy levels and severance payments, redeployment issues, and levels of pay related to the new jobs that existing NGA staff were expected to do.

The deputy editor openly admitted that the management's approach was to pressure the unions into an early settlement, but this idea was brought to a halt when the managing director retired. However, it seems clear from the organizations' point of view that the negotiations were not about the jobs or even the people. Decisions concerning these factors had already been made. The argument was over money, and some time was spent discussing where remaining workers would go, but the job definitions were decided long before. Neither did the deputy editor make any attempt to suggest that there was consultation. It was a management imposed programme, the newspaper was only concerned about discussing the terms under which the programme would be implemented. Most important, was that it would be compatible with

the editorial and advertising staff, any consultation with
production staff was dismissed as irrelevant:

> I don't think it's necessary to consult the production staff about that....what I was
> concerned about was the knowledge and experience of, in particular, journalistic
> departments and what we did on screen matched what they wanted to do....what I want to
> do is to build a system that works for journalists, I don't want a journalist to be
> sitting down doing compositors work.... (Deputy Editor, Hermes).

The purpose of direct entry is to use editorial staff, with the
aid of computers to do the work of composing staff, therefore the
notion of consulting a compositor is irrelevant. The managers had
reversed the early scenario of NGA supremacy as described by
Cockburn (1983) and Martin (1981).

Mercury did not adopt a macho management stance like that of
Hermes and did not have the same industrial relations problems
as those at Hermes. However, similarities exist in the
newspapers' wish to break trade union control over working
practices. The production manager at Mercury reported 'a change
in attitude' of the union employees and also commented that the
trade unions '....were not the force that they had been in the
past....' These factors undoubtedly made a good environment for
the change to direct entry to take place, and like Hermes,
Mercury had every intention of capitalizing on the divide between
the NGA and NUJ to marginalize the NGA. The negotiations did
eventually include both unions, and the production manager
claimed that the initial concern was unfounded and he was quite
happy with both unions being present.

Consultation procedure

On the whole, the negotiations seemed to be handled clumsily in
Hermes. Arrangements were made for the direct entry announcement,
and these were planned simultaneously, but separately, for both
the NUJ and the NGA with SOGAT. This was an early attempt to
simply ignore the Accord and adopt a 'divide and rule' strategy.
However, the basis of the accord was that everything should be
done jointly, so the NUJ members joined the NGA at their
presentation. The announcement was accompanied by a management
publication *A Presentation to Staff in the Composing Room*, which
detailed the changes and the options available to production
workers. These were voluntary severance, early retirement,
redeployment, or a 'wait and see option' which involved staying
with the company until the new system was established, and then
deciding. From a management point of view this proved to be
effective, many believing that this statement of intent was not
negotiable. The NGA FOC at Hermes, however, took the statement
as a starting point for negotiations.

The start of negotiations was marked by the use of the
Newspaper Society, whose negotiators refused to conduct talks on
direct entry and redundancies, claiming that these were unrelated
issues. They also objected strongly to the concept of a joint

agreement under the Accord. The NGA FOC was very unhappy at this approach, and industrial relations became difficult. Shortly afterwards, the managing director took early retirement.[4] Following the replacement of the managing director, the Newspaper Society was removed from the negotiating table. This transition period involved some months in which no progress was made, and after a new managing director was appointed, there was a further one month period in which he familiarized himself with his new environment.

While these problems were not evident in Mercury, trade unions in both newspapers won concessions on the number of redundancies, the levels of severance pay and redeployment terms. The negotiation was dominated by financial issues, and the NGA failed to offer any alternative scenarios for work reorganization, although they were given little formal opportunity to do so. However, the negotiation, and the accord did test union solidarity and did emerge as a battle between management and unions. In this kind of relationship, union involvement in managerial decisions was extremely unlikely.

Trade union approaches

The NGA was in a weak position, the impetus for new technology introduction having built up over a period of years. Management had been forced to compromise by accommodating the existing demarcations, even where the workers seemed unsuitable in previous changes (Cockburn, 1983). They were no longer prepared to do so. The NUJ were in a much stronger position. For their members, the changes were likely to be beneficial. The 'Accord' between the NGA and NUJ provided that no issue be discussed unless both unions were present (see *NGA/NUJ Agreement, Direct Input in Provincial Newspapers*). The management of both newspapers found this distasteful, since such an alliance had the potential of bringing the newspapers to a complete standstill. However, eventually it was accepted, and in negotiations issues such as severance payments and redundancy numbers, as well as training, health and safety, and ergonomic issues were discussed. Discussions about alternative scenarios for job design and work organization were not evident.

The NUJ FOC at Hermes explained that there was a strong feeling that the NGA was attempting to secure an agreement that retained a strong chapel, and the managers were trying to weaken the chapel. The pursuance of such political aims was happening anyway, but the prospect of direct entry offered an ideal opportunity to drive them home. The newspaper actually achieved two separate agreements, which were essentially the same, but separated because of management concern about the power of two unions bonded by one agreement.

The negotiations at the newspapers, particularly Hermes, seem governed by a hidden political agenda, which achieved its goal in wresting control over production from the NGA, but developed a common interest between the NUJ and NGA.

Summary - newspaper studies

The content of the newspaper negotiations was limited by management, the main decisions had already been made before there was any announcement to the trade unions. The main concern of the trade unions was the retention of staff and maximizing severance compensation. Again, there was little opportunity to attempt to influence work organization, but what the NGA at Mercury did achieve was a considerable reduction in the agreed number of redundancies and also played an important role in persuading the newspaper that readers should be retained.[5]

Although the complexity of the industrial relations problems stand out at Hermes, both the newspapers had a number of similarities in their approach to negotiations. Both were opposed to joint negotiations, and neither included the trade unions in any form of decision taking, prior to announcing plans for change. The weakening position of the NGA left them in a poor bargaining position. Managers had no intention to invite trade union views on new technology and work organization/design, but neither did the NGA attempt to present any alternative possibilities as challenges to the management's prearranged strategies. Managers relied on conflict between the NUJ and NGA as a means of diverting attention away from the management, and thereby reducing levels of opposition. Instead the opposite happened and the NUJ and NGA joined under an agreement to ensure that neither party was disadvantaged.

Confirming the hypotheses, the trade union had few ideas for tackling qualitative issues, but this was against a background of considerable management opposition, and a management agenda for negotiation which was purely quantitative.

This consistency of approach was not true of the clerical studies which vary in their approaches, from both a management and union point of view.

Office studies

The negative attitudes of Apollo's (the aerospace company) managers towards the trade union's organization, causes difficulties for the inclusion of the trade union in discussions. However, here the trade union is effective in making a case for inclusion in new technology discussions. Their lack of success in gaining access is largely due to managerial intransigence. Achilles (the shoe company), unlike Apollo, had both a management and a union who failed to see the importance of involvement in work organization issues.

Managerial attitudes

The personnel manager at Apollo claimed good relations with the trade unions on the site. He explained the way in which trade unions were given company finance for training their officers, and how union representatives were sent to see the technology in operation. In these ways, the company claimed that there was rarely any objection to introducing new technology because the 'ignorance factor' was eliminated. Much of the education seemed to have been directed at removing 'the myth' that technology caused a loss of jobs. Involvement in job content issues seemed negligible, and when questioned on this the personnel manager would refer constantly to the union's involvement in job grading issues. His confusion of the two could have been either because he did not make a distinction between the issues, or because job gradings were corporate personnel issues and content was generally a departmental issue.

While the personnel manager presented a picture of a relatively harmonious working environment, this was rapidly shattered in the accounts department. The financial accountant himself displayed an uncompromising assertive style. He explained that some five or six years ago (interviews were conducted in 1988) the trade unions were advised that new technology would be coming into the business, and changes in jobs should be expected. As at Zeus the union had agreed that only technology to a level that the staff could cope with would be acceptable. Trade unions were not consulted on the introduction of new technology but were informed of its impending arrival. Again this was done in the proactive style of Zeus, preparing the workforce in advance. When the new technology was introduced, APEX reacted by applying for pay increases for payroll staff outside the agreement made with the whole company.[6] However, the financial accountant saw this attempt as something of a challenge to his managerial authority: '....payroll was a classic and I threatened to dismiss four of the staff and told them on the Friday, they had until the Monday to make their minds up. If they didn't work in accordance with the new system and the new technology they would have to consider themselves....off payroll.' By the Monday, the financial accountant explained, the union had realized they were wrong and opted to keep their jobs. Not surprisingly, problems of this nature have not recurred in the department.

Hence, what initially appeared to be good relations at Apollo were later revealed, by further examination, to be the opposite, similar problems were experienced in studies by Batstone *et. al.* (1987). In some departments, trade unions were viewed with contempt, and problems approached reactively and aggressively. Achilles' managers had difficulty seeing any constructive role for trade unions (see also Wainwright and Francis (1986).

The manager at Achilles, looking after the electronic point of sale developments, claimed that few of the shops involved had union membership so it seemed inappropriate to consult the trade union. There was however, extensive consultation of shop workers and managers which was seen as essential for achieving efficient operation of the system.

In the accounts department the credit manager adopted a similar 'need to know' criteria for consultation. He suggested that the recent changes did not require consultation because there were to be no job losses. He then expanded his terms of reference to include deskilling: 'If we thought there was any likelihood of current staff, and I emphasize current staff, being, or having their jobs deskilled, then of course we would involve the union as a company, we always have done.'

His reference to current staff is significant, since plans were afoot to upgrade existing staff, and possibly use new staff to handle the mundane work of the department. Apparently there had been no consultation with the staff or the union on this changing composition of the work in the department because existing staff were not to be deskilled and no job losses were anticipated. While the new technology had not yet arrived, the main decisions had almost been finalized, again without any reference to the union or staff. I was not permitted to question members of staff on new technology issues because management had yet to inform them of the precise details of the changes. Most were aware of the impending changes but the knowledge they had was based on hearsay and rumour.

Consultation procedure

The financial accountant at Apollo claimed that the union were involved in some discussions about new work organization arrangements. However, the use of different union representatives at successive meetings brought confusion to the discussions, and none of the union representatives knew how far the changes had been discussed. Once again applying managerial prerogative, the accountant deemed it necessary to cast aside the consultation and make the decision without the union. The personnel department protested strongly but ultimately had to follow the accountant's decision.

The management accountant, working alongside the financial accountant but with fewer staff responsibilities, did not see that workers should be particularly involved in any decisions about the new technology or the jobs. He argued that if any member of staff had an important point to make regarding the new systems they should come and discuss their ideas with management. This was an unrealistic attitude given the problems associated with consultation that had occurred in the department, the staff are extremely unlikely to want to approach management on any issue.

However, not every department held the same views as those in the accounts department. In the manufacturing services division there was a considerable contrast with the financial accountants view:

> What I'm saying is those guys out there know a darn sight more than us on what is happening out there, therefore they've got some good ideas on how to improve....I have a firm view that a lot of those people have got a lot to offer if only we care to ask them, we haven't done to date (Operations Manager, MSG).

The claim of systems analysts that they would discuss implementation of systems with all levels of staff, was not borne out by the trade union or the staff, who pointed out that analysts rarely consult below managerial levels. Some of the analysts were not concerned with the wishes of the staff, others would be instructed by managers that they were able to give full information, and that talking to staff was unnecessary. There was also a tendency, where advice took place after the implementation of technology, to go to staff directly and not through the trade unions, which also occurred in studies by Storey (1987). It was an area of considerable contention. The local branch of APEX was clearly aware of the issues but having problems convincing the management that new machinery introduction was a negotiable part of the job.

Trade union consultation at Apollo was not seen as relevant in some departments, while in others managers claimed that they recognized the importance and the benefits of this. Yet in all departments the evidence suggested that little practical consultation was taking place.

At Achilles, the management information systems manager suggested that the standard approach was for the section manager to consult with the trade unions initially, and then update as time went on. However, his was one of several perspectives on what the level of consultation should be. Another systems manager suggested there was no guarantee of consultation but that '....unions are involved when there is a need.'

The procedure was for management to inform the branch chairman of any major issues, and the union would decide on its approach after consulting its members. However, in the case of a change of an employees' contract, the union must also be informed but will only take action if the person is unhappy and has a reasonable complaint. There was, apparently, a procedure for consultation about new developments in areas such as new technology. This seemed to be the kind of positive approach and interaction that was required. However, the procedure generally failed to operate, and the branch chairman's attitude certainly did not help: 'I've got plenty to do without worrying about going to another meeting once a month.' Clearly the problem at Achilles was a lack of interest on the part of both managers and union representatives, unlike Apollo, where it seemed to be located with managers.

Trade union approaches

For Apollo, as at Zeus, another function of keeping the workforce
well informed was propagandist, since there was naturally much
suspicion when the company introduced new technology.
Consultation has the effect of acclimatizing the workforce to the
changes rather than creating anxiety and suspicion on
implementation.[7] However, there was evidence of significant
employee consultation in the changes, and recognition that there
was a lot of room for improvement. However, at departmental level
there was not a great deal of evidence of the trade union trying
to influence issues, beyond attaining higher financial rewards.
Nevertheless, the trade union on site was well organized and
aware of the effects of new technology in the business.

The main clerical trade union was APEX which had a chairman and
secretary on site. There was also a separate negotiating body to
whom problems are referred. Union representatives recalled a
redundancy programme in 1982 when 149 staff were lost, all were
volunteers, but there was continuing concern that the staff were
being reduced through natural wastage all the time. Both of these
developments were thought to be a result of new technology
introductions.

Another complaint, directly connected to the use of new
technology, concerned the way the training was conducted. The
trade union claimed that the benefits of new technology were
often lost because the management failed to provide sufficient
training for its use. Actual examples included the new MANMAN
system (dealing with stock control) where workers were given only
one day per week across the period of a few weeks. This was
considered too little to gain a fluent understanding of the
systems. In other areas, operators were being released after just
two days training onto systems they did not fully understand.
Even 'on the job' training was considered insufficient because
no one person had enough knowledge to troubleshoot the system
rapidly. Staff members complained that even system analysts would
be constantly referring to manuals and the system suppliers to
solve particular problems.

The union secretary, a wordprocessing supervisor, advised that
while the use of wordprocessors had been discussed, in relation
to its effect on typists (and even here the equipment to be
purchased was discussed up to a point), concern arose because
with their introduction, managers and engineers began
wordprocessing their own reports. This caused considerable
anxiety amongst the word processor operators.

The APEX secretary commented:

....we have sole bargaining rights for the clerical workforce, that's all we have, at
one time the company used to keep us informed of everything that was happening nowadays
they just seem to ignore us, they just bring it in and say - it's here what are you
going to do about it?

The APEX chairman was keen on making a case for the inclusion of the union in detailed discussions over work organization, and indicated an understanding of the concept of 'tacit skill': 'There is a lot of knowledge amongst the workforce, which, if management is willing to take notice of it, would make a job a lot easier....there are certain things that they are not aware of which the workforce is aware of.'

The chairman also felt that there was too much emphasis on fitting the jobs around the technology, rather than buying the technology with the people and jobs in mind. This he felt was possibly based on the idea that there would be an automatic adverse reaction to new technology. APEX had succeeded in drawing up a new technology agreement, with the management, based on issues such as consultation, declarations that no job loss would be as a result of new technology, and the redeployment of personnel. The latter being conceived primarily for the older workers. However, many of the agreements had been broken by the company, with the exception of the health and safety aspects associated with new technology, and this had promoted a good deal of distrust and sour relations.

APEX was well aware of what was happening in the organization, and was armed with constructive views on the implementation of new technology, demonstrated by their establishment of a new technology agreement. However, as in office studies by Batstone *et. al.*(1987), the company did not seem committed enough to want to hear those ideas. Unlike some of the other case studies there was no preoccupation with financial matters at company level, although this was evident at departmental level.

The lack of awareness amongst union representatives does not seem to be the main problem at Apollo, but the willingness of management to listen to union views certainly is. Negative attitudes of the union's ability to be of practical assistance, restricted their inclusion in discussions. Even where there was recognition of the advantages that the trade union could bring to the company, limited consultation took place in practice.

For the clerical staff at Achilles the trade union was also APEX, who here, did not seem nearly as well organized as the union at Apollo, and certainly less aware of job organization issues. The concern of the management and the union in the distribution division focused on the problem of the impending loss of staff. They agreed that as a current member of staff left they would be replaced by a temporary employee, who would be easy to remove when the time came. When the new systems were implemented, the trade union did not seem too concerned with the job definitions themselves, although clearly the company was aware that employees were anxious about the changes. A good deal of redesigning and redeploying was required. It seems that the discussions to arrive at this took place, not amongst the trade union and the employees at all, but with higher levels of management where the decisions then filtered down to lower managerial levels and were subsequently implemented.

For the trade union, work organization issues were of no particular relevance. The branch chairman of APEX saw little gain in approaching the staff on job design issues, and saw issues such as job loss, severance payments, and health and safety issues, as the important features of the union's work. This was despite the lead taken by the national organization of APEX in new technology and job design issues (APEX, 1985; 1980).

As was so often the case, the trade union tended to make the association of new technology with job loss, and focus on this to the exclusion of other important areas. Although in both this study and others, managers seem to have been allowed to set the agenda for consultation, excluding unions, but in many cases the unions have not challenged this wisdom. In Achilles, APEX adopted a very narrow view. Deskilling was recognized as being an issue, but not one the union had to worry about, since even if deskilling did take place gradings were guaranteed. There was no recognition of the, potentially profound, non quantitative changes that might be taking place in work content, and the possible adverse effects that this may have for the workforce. Health and safety was an issue that the union was particularly concerned about, and close attention was paid to ensuring the company was not in breach of any ruling that was in force. But again health and safety, although qualitative in nature, is more tangible than issues of job design, and is perceived as particularly important.

Summary - office studies

Apollo had made considerable inroads into consultation with management over new technology issues. These dealt with work organization, as well as health and safety issues and attempts to prevent job loss. However, the problems that were encountered here were a result of management's approach which was negative in the face of a relatively informed, and potentially positive, union approach to new technology. There were examples of poor consultation in the organization as a whole, but also particular managers seemed determined to impose their own individual style which often included the exclusion of trade unions.

Achilles, on the other hand, lacked both a committed management and knowledgeable trade union. Apathy, and lack of awareness in the trade union, had meant that managers had rarely been challenged on any issue connected with new technology introduction, outside the employment and pay level questions. Managers were not as aggressive as some of those in Apollo, and they may have been more susceptible to union pressures had a case been presented.

In both cases, the trade union for clerical workers was APEX who has made important contributions to the debate about the role of trade unions in new technological change processes (APEX, 1980) and job design issues (APEX, 1985). The evidence suggests that in the case of Achilles this information had failed to filter down to the local union organization.

The evidence is consistent with the hypothesis in the case of Achilles, but in the case of Apollo it is clear that a potentially effective trade union, fails to establish effective dialogue due to adversarial management approaches.

Occupational mergers and traditional demarcations

As we have seen, new technology can result in deskilled detailed tasks or more highly skilled work, largely dependent on the way managers decide to organize both the technology and work. However, it became clear in preliminary pilot studies, and in an analysis of existing case study material, that neither deskilling nor the addition of skills could be seen as autonomous. This was confirmed in chapter five, when the hypothesis that cases of reskilling and deskilling would exist together, was supported by evidence from empirical studies (eg. Batstone *et. al.*, 1987). In some cases it was found that deskilling and reskilling would take place in the same department.

There is a further important dimension, that leads on from this notion of a multiplicity of skill levels as a result of new technology introduction. If new skills and new technology have the ability to reskill and expand the skill repertoire of workers, there is a real possibility of conflict developing as a result. Expanded skill repertoires of some workers, may begin to encroach on the work areas of others and breach established demarcations. Where clear skill demarcations have been established, trade union conflicts may result which may be identified, both within trade unions, and between trade unions. The latter may occur in cases where the union represents a variety of skill levels.

In some cases of new technology change there may also be a blurring of the boundaries between some jobs. In some examples this is more pronounced, and the nature of work changes completely. In these cases there may be conflict within trade unions, as one union sees the members of other unions undertaking work that was previously exclusively their domain.[8] Similarly, within one union, conflicts can occur if the members of the union are within a skill hierarchy, and attempts are made to add tasks, previously held within the skilled demarcation, to those whose jobs are semi-skilled or unskilled. Elsewhere, the conflict may not involve the trade union, but because new technology can be used as a means of upgrading people to take on new responsibilities, and of controlling processes and procedures, there is a real possibility of conflict with immediate superiors who fulfilled these tasks previously. Hence, this hypothesis suggests that conflicts between groups of workers may result where new technology is implemented, demarcations are breached and jobs are changed.

If this were the case then there are important implications for managerial and trade union approaches to the definition of jobs. Management decisions about job design need to consider the likely conflicts that may be stirred up, and trade unions have to look to their relations with other unions and the relations within their own unions.

Metalworking studies

In both of the metalworking studies, the main areas in which occupational mergers were taking place, were where CNC operators were taking on programming tasks. In Zeus, maintenance workers exchanged electrical and mechanical knowledge, and there is also some evidence of supervisory authority being passed down the hierarchy.

Demarcation lines

In the maintenance area of Zeus, where the possibility of divisions occurring between unions seemed likely, the personnel manager denied that there were any problems of the unions attempting to maintain their established demarcations. However, the trade unions had to consider the threat of the use of contract labour should they introduce any serious opposition, and from this point of view acceptance of change was more of a compromise than evidence of a consensus. The jobs had been changed here so that electricians could handle the major part of a fitter's job, and the fitters a similar percentage of an electrician's job.[9] The personnel manager also claimed that the fitters and electricians could clearly see for themselves that the strict mechanical/electrical division could no longer be maintained with some forms of new technology that blurred the traditional divide between the two trades (see also Jones, 1988).

The Zeus Era plant also offered opportunities to investigate the potential threat that senior CNC operators posed to production engineers. However, at Zeus, the CNC operator is semi-skilled unlike those at Artemis, who are all skilled (Jones, 1982; Jones, 1983; Wilson and Buchanan, 1988; Wilkinson, 1983).

Also at Zeus, another area where demarcation lines were challenged concerns supervisory staff. Reductions in supervisory staff by one third have taken place but remaining supervisors have themselves had new managerial responsibilities passed to them. Here, the production operations manager felt there was little problem of supervisors becoming anxious about the security of their jobs. However, concerns were expressed in other areas, and one production manager suggested that some supervisory staff felt threatened by the use of 'improvement teams.' There was also talk of producing a new grade of supervisor called section leaders who would take on wider responsibility. Supervisors expressed concern because of the recent reduction in supervisory staff. Judging by the major ways in which Zeus was changing at each subsequent visit, it seemed possible for there to be further reorganization, despite the substantial reorganization that had already taken place.

At Zeus, there are three areas in which possible conflicts can be identified. The maintenance area disputes over the exchange of electrical and mechanical skills, the conflict over CNC operator programming, and the handing down to lower grades of supervisory responsibility. Artemis was also vulnerable to conflicts over divisions drawn between CNC and conventional workers, and between CNC operators and existing programmers.

For Artemis changing the grading bonus system placed CNC operators at the top of the hierarchy, which upset the conventional operators who considered themselves more highly skilled. The shop steward argued along the same lines as his equivalent at Zeus; CNC was button pushing and a set of very simple tasks. In terms of CNC operation, it was apparent that in both companies there were potential divisions between CNC and conventional operators.

As at Zeus, the other area of interest at Artemis, was the allocation of CNC programming work. Unlike Zeus, no evidence emerged to suggest that existing programmers saw operators as a threat. Managerial attitudes were not entirely in favour of securing a full programming role for operators and as a result such a role was unlikely to materialize.

Demarcation conflicts

One area of union intervention was based on concern that no one was embarrassed by not being able to handle the additional skills, and particularly, that their skills were not acquired by semi-skilled staff who might then begin to handle their own maintenance. This was an AEU initiative and demonstrates that while the skilled maintenance workers themselves wanted to learn new skills, they also wanted to ensure that nobody else was learning their skills. Similarly, the EETPU were happy to learn the mechanical work, but were clearly unhappy about the reciprocal arrangement with the AEU members.

As the majority trade union in the company with negotiating rights, the AEU was able to exercise significant influence over the relatively few EETPU members. A similar scenario was experienced by Jones (1982) between the AUEW and TASS unions. However, there clearly were problems between the AEU and EETPU in the early stages of negotiation and these were revealed by the personnel manager despite his earlier claims:

....there has been a lot of animosity behind the scenes from electricians who argued that they were the craftsmen of the future which caused a lot of dismay with mechanical craftsmen in that they believed it to be true, they wanted a way in which they could get the commanding heights of electrical maintenance without having an inter union dispute.

The fitters that accepted training, rapidly reached a high level of competence in electrical maintenance, and at this stage the electrical craftsmen started to try to limit where non EETPU workers could go on the basis of safety considerations. Having already been outvoted by the AEU on the general issue of adding electrical tasks to mechanical craftsmen's jobs, the electricians then suffered a further defeat when the factory inspector ruled that the training made the fitters fully competent persons for electrical work in the factory. Nevertheless, the EEPTU persisted arguing that there should be demarcations to prevent AEU workers from opening the electrical panels on CNC, and other similar machines. Had this amendment been accepted, it would have caused severe problems for fault diagnosis, and would have largely defeated the object of the multi-skilling training. Management and AEU, in concert with each other, rejected the EETPU proposal. A number of strikes followed by the EETPU craftsmen, but the change was eventually accepted when it was clear the company was not willing to compromise. The AEU shop steward summarized the problem as the electricians wanting to learn the maintenance function, but not wanting the fitters to learn an electrical function. There was little sympathy for the EETPU from the AEU. However, in context, the reaction of the electricians is easier to understand, since there are many more fitters than electricians, and another group of factory service personnel[10] were also taking on electrical skills.

Interviews with a number of machinists and supervisors in machining areas, suggested that there was a degree of conflict between CNC operators and conventional machinists. This was based on the myth that CNC operators were only button pushers and did not deserve their positions at the top of the semi-skilled grading (see also Wilson and Buchanan, 1988 and Batstone *et. al.*, 1987). The CNC operators strongly disagreed, although acknowledging that little skill was required if everything ran smoothly, their skill was only really evident when there were problems:

> I think they sort of look at you and when it's running okay you just press a button. I mean <u>that is all you do</u>, you're just a button pusher while it's running okay. But it's when things come up wrong and they don't see that....Either because they don't want to see it or they are just not bothered. They only see the side when you press a button and they think that's all you have to do, you can only glamorize it so much.... (CNC Operator, Grade B - his emphasis).

Much of the conventional operators' argument was clearly based on misinformation, but similar claims were voiced by staff working alongside CNC operators, and perhaps more importantly, by trade union stewards who argued that CNC machinery promoted a process of deskilling while the conventional machining function required high levels of skill and craftsmanship (see Braverman, 1974; chapter two).

One of the supervisors suggested that programming by operators would not threaten all of the existing programmers who were production engineers. Some, he claimed, would be quite happy in the knowledge that operators were limited in their ability. However, one of the operators working on a Sharman machining centre wanted the opportunity to learn more but had been unable to secure the training time. In this he felt the production engineers had played a part: '....they [the managers] wanted us to do everything but the production engineers sort of stood their ground a bit there. It wasn't actually a conflict....'

The dilemma that these operators faced, was that without the formal training, the only way they can learn is informally from the production engineers. But there were limits to the extent that production engineers were willing to spend time with them. One production engineer interviewed, claimed he had no objections in principle to operator programming but doubted their ability. However, there was clearly an undercurrent of concern about his tasks being taken away:

> ....bear in mind that all our operators are semi-skilled, we haven't got a skilled operator, so we've got some good guys and some not so good guys, and they all work the machines to the level of their incompetence if you like. But we've got one or two good guys on the lathes who we allow to have the head on programming (Production Engineer, Zeus Era).

However, the 'one or two good guys' were programming for the production of pulleys on simple two axes machines. He doubted the ability of operators on anything more complicated, and questioned the enthusiasm of operators to want to taking on the responsibility in planning:

> They do some programming, some of them, but it's dangerous, you can't call them programmers. When you've turned lots of round parts like that and you've been doing them for a year and you've made varieties on that thing, different varieties, different diameters, different widths and you understand your machine. It's not very difficult to write a program for this part once you understand this part and this is basically what we're doing out there. To show the limitations we came in one day and we had a load of bits on the floor all sort of smashed up, the tooling a right mess....the problem isn't just programming it's tooling, knowledge of tooling, knowledge of the program, knowledge of cutting feeds and speeds, methods - what shall I do first, face it, turn it....?

To go to operator programming the production engineer felt that training like his, a five year apprenticeship, was required as well as a thorough understanding of the specific machinery. The company could not afford expensive errors with tools and equipment. There was concern that both the AEU and the management were promoting operator programming, although whilst the management clearly wanted an expansion in skills, there was little evidence of local trade union knowledge or interest of the area. The main argument of the production engineer was that of knowledge and ability, a secondary argument was a question of

roles in the organization which can be read as demarcations:
'....we have the guy on the machine to make bits, if we were to
have him writing all his own programs....I don't know who would
make the bits.'

The demarcation issues did not figure with the representatives
of either the AEU or the MSF, representing operators and
production engineers respectively. Although it was felt it may
become an issue, and the production engineer eventually did admit
that the recent redundancy programme had sharpened most peoples'
awareness of their job demarcations. In the foreseeable future,
this production engineer made it clear that he would not be
willing to train an operator in programming.

The production operations manager agreed that it would not be
easy to train operators to program on five axes machines, such
as the Sharman machining centres. However, there was obviously
awareness of the demarcation issues and the manager suggested,
not only would it be beneficial for the operators to have
knowledge of programming, but he extended this to include
maintenance also.

Retaining the maintenance function demarcations was a
particular concern of the skilled craftsmen at Zeus. Within the
AEU the division exists between the skilled and the non skilled
staff. For the skilled workers there is concern that the semi-
skilled will begin to take over tasks that they recognize as
theirs. Another encroachment on their skills, that was also
resented, was the increasing use of degree or diploma qualified
staff who had not been through the traditional apprentice
training procedure. There was common ground between the AEU and
EETPU on these issues, and attempts to hand electrical or
mechanical maintenance tasks to semi-skilled machinists would
certainly meet with vehement opposition from the skilled union
members, as would any attempt to use sub contracted labour:
'....they would still guard their maintenance skilled status very
jealously if we tried to encroach into it significantly with what
they see as non skilled men' (Production Operations Manager,
Era).

Clear conflicts had developed as a result of the multi-skilling
arrangements involving the maintenance function. Potential
conflicts were evident between existing programmers and
production engineers competing for programming work. Another area
where challenges were likely, was the passing of maintenance
skills to shopfloor workers. Artemis displays considerably less
evidence of conflicts.

Having made the decisions about work organization, some
managers at Artemis clearly felt that some degree of
justification was required, to offset the unrest between CNC and
conventional operators. The industrial engineering manager saw
it as managements' duty to inform the workforce that CNC was not
as simple as it may seem, and required a background in
conventional skills as well as the additional CNC skills. This

education was particularly important to ensure that the new grading structure, and the new bonus scheme, could be ushered in. For the personnel manager, the problems occurred: 'One, because a proportion of them would lose bonus earnings in the averaging out across the shop, and secondly, we were turning over the perceived level of skills.'

However, the personnel manager explained that although still a problem, as people progressed through the module training programme they became more aware of what was really involved in CNC operation. This was because the higher levels of training involved aspects of CNC work, and workers began to appreciate from actual first hand experience of doing similar work, that the CNC operators were not only button pushers.

There were many staff who had always worked on conventional machines, and either did not want to train for higher gradings, or had been dissuaded by the company from doing so. In the company's view some workers were incapable of going past grade one. For this group there would always be a view that the skills required for CNC, were inferior to those required for conventional operation. However, the company had future plans to have a workforce of entirely multi-skilled operators, capable of using both CNC and conventional lathes.

The senior shop steward in the machine shop at Artemis, pointed out, that he felt additional skills had extended the knowledge of the workers, making them far more indispensable than before. The objective of the reorganization and comprehensive training programme was an attempt to alleviate a shortage of programmers, and increase the flexibility of the staff as a means of dealing with peaks and troughs in the business, requiring staff to transfer from one operation to another rapidly. Indeed, Artemis should offer a good example of an organization that has witnessed conflict between two groups of workers, due to its enhancement of operators' roles that encroach on the work roles of production engineers. The industrial engineering manager explained that the emulation of programmers' work was entirely by design:

> ....to accomplish all the activities in technician grade three you are getting quite close to a production engineer and that was the concept we wanted to follow, it wasn't something that we were worried about or frightened about, it was something we wanted to actively encourage. Our basic train of thought was that the man would not necessarily be occupied all of the time, if an CNC machine is working away, loads of tools, a number of pallets will change automatically and the operator isn't physically involved in the cutting process. At that time he's not very important but when something starts to go wrong you want someone there to do something about that very quickly.

Here then the programmer is not really threatened, the limited extent to which the technician gets involved is to alleviate a problem when no programmer is present. Limited amending tasks are a common occurrence (Jones, 1982; Burnes, 1989; Wilkinson, 1983). So despite the rhetoric of both the production director, and manufacturing manager, about lifting the technician grade threes to the level of programmers, the practice seems quite different.

In confirmation of this, the personnel manager saw no possibility of such a shift in responsibilities:

> Certainly at this present point in time I can't see a change because the work is becoming more complex both in terms of technology and in terms of a range of products. So what we need, and we're going back to basic principles of management, whereby there are people who plan and organize, and those people who execute. Those people who execute are the machine shop technicians (Personnel Manager, Artemis).

Any aspects of planning, or bordering on planning, that the technicians might handle are restricted to minor amendments and changes, should the system go wrong. Interestingly, the personnel manager demonstrated that not only was the job of production engineer a specialist one, preventing operators from reaching that level of competence, but also that a grade three operator would have more to lose, since the pay received, when overtime and bonus payments are included, is more than that of a production engineer (Jones, 1982; Burnes, 1989). However, the real difference in opinion between personnel manager and industrial managers, was probably due to the personnel managers more acute awareness of the problems that would develop should production engineers be challenged for their programming functions.

The operators themselves felt that the switch to production engineer was equally unlikely. One grade three operator complained that some grade three staff are still restricted to conventional machines, despite the policy of using this level of operator on CNC, the training modules seemed very arbitrary:

> The grading system leaves a lot to be desired really. It's not really fair, they give it to some people and don't give it to others but there is no reason to say why. They laid out a module, some people don't even know half of them....There is no basis for making people a higher grade really, no straight format. They don't say like, when you learn that you'll be a grade three, it doesn't work like that (Machine shop technician grade three, Artemis).

Based on this information, the grading system does not seem capable of carrying someone to programmer from the operators' ranks. In the current environment at Artemis, it seems that the programmers have little to fear from the expansion in operators' jobs.

Summary - metalworking studies

Clearly, conflicts of interest that may be generated by new technology do not exist only between managers and unions, but within and between unions, and between workers. This friction may challenge a traditional hierarchy established in a firm. New technology in the engineering companies, had the effect of blurring the traditional demarcations between the jobs of skilled maintenance workers, skilled and semi-skilled workers, and CNC and conventional operators. For the latter group the problem was more one of the perceived level of skills in the workforce. The implications for defining jobs, point to the problems managers may experience in shifting job boundaries freely, without careful prior consultation.

In Artemis, there was little evidence of conflicts occurring, although the potential for this existed. Zeus provides more interesting examples, particularly in the multi-skilling of the maintenance function. In line with findings of Dodgson and Martin (1987), Jones and Rose (1987) and Jones (1988), trade unions were willing to accept the benefits of new technology changes to their own jobs, and have little regard for other resulting job changes. The AEU recognized the advantages of adding electrical skills to their jobs, but were unsympathetic towards the EETPU who objected. The same writers have noted that such changes can place managers in favourable positions to achieve their goals. Here again, the example in Zeus is where managers used the opposition toward the EETPU to win their case. However, on other issues, such as the question of handing maintenance tasks to operators the EETPU and AEU would provide a united front in opposition against another part of the AEU. This tendency for alliances and conflicts between, and within unions suggests an instability that may leave trade unions highly vulnerable allowing managers to use such weakness to their advantage.

The evidence suggests support of the hypothesis in these studies. There are actual conflicts between trade unions and potentially other problems based on the merging of occupations which contradicts existing demarcations.

In the newspaper studies a history of traditional conflicts suggested that there would be acute divides between workers, and between the trade unions involved.

Newspaper studies

In the newspapers, demarcations were traditionally some of the strongest. The NGA pre-entry closed shop, ensured that production demarcations were sustained and this lasted until the direct entry changes arrived. However, initial skirmishes between the NUJ and NGA, notably the case of Portsmouth and Sunderland Newspapers (Industrial Relations Review and Report, 1985) and the Wolverhampton Express and Star (NGA, undated), perhaps made the unions realize that their disunity left them vulnerable. Instead of there being conflicts over inputting work that journalists were to take from compositors, NGA workers accepted that they would lose their old jobs, and NUJ members assisted them in negotiating the terms on which the changes would take place, under a temporary alliance called the Accord.

Demarcation lines

Divisions between the trade unions in the printing industry are well known, and new technology's ability to simplify production jobs, and enhance editorial work in the industry, provides the conditions for greater rivalry. However, the inevitability of the changes, and the knowledge that each of the main unions could hurt the other, meant the NUJ and NGA decided to form an alliance. SOGAT '82 was the industries third main union covering semi-skilled and unskilled workers.[11] Neither newspaper had many SOGAT '82 members, those that were members tended to be in areas such as the press and generally in the production departments. For example, at Hermes there were nine SOGAT members, six of these were employed on the press, the remaining three were involved in publishing.

SOGAT members stood to gain from the changes precisely because the NGA monopoly on skilled work was about to be broken. The production manager at Mercury was quick to point out that, before the change to direct entry[12], there were many jobs that still could not be interchanged because of the strict demarcation lines, however:

....in essence they do the same bloody job....but because one happens to hold one card and one holds another card one gets 87.5 per cent of the others salary....He does virtually the same job and he could quite easily do the same job, he's intelligent enough to do it (Production Manager, Mercury).

Hence the NGA had maintained strict demarcations, and it was impossible for the company to employ a non NGA person in skilled production work. It was necessary for skilled production personnel to be members of the union if they wanted to work in the printing industry (Cockburn, 1983). The pre-entry closed shop[13] was defeated alongside the breaking of demarcations in the industry with the introduction of direct entry technology. New staff now have the choice of which union they wish to join, and the choice of not joining a union at all. The NGA, which was previously exclusively skilled, has begun to recruit new workers in the production area irrespective of the level of skill. The firm also seemed encouraged by the prospect of the likely wider membership of SOGAT who had, compared to the NGA, been a relatively unproblematic union. The production manager at Hermes was very insistent that SOGAT should be the appropriate trade union for the advertising area:

....with direct entry going into advertising we reckon SOGAT, as the appropriate trade union, should be advertising for people that want to join the trade union, at the moment they're not (Production Manager, Hermes).

This production manager's eagerness to have SOGAT as the main union, became clear when the NGA FOC revealed how weak and ineffectual the trade union actually was at the newspaper. For example, he pointed out that at a recent meeting nobody turned up. This evidence, and similar indications in chapter five, suggests that managers wanted to see their companies free of trade unions, or having only a token representation. One of the ways that this could be achieved was to exploit the rift between the NUJ and NGA, diverting attention away from changes that were actually being made.

Demarcation conflicts

The NUJ and NGA have had a difficult history, when direct entry changes were in their early stages in the provincial press, both the NUJ and NGA made agreements with newspapers excluding the other union.[14] Interviews with management, before negotiations had begun, suggested that they would capitalize on this relationship employing tactics of 'divide and rule.' Eventually, however, the unions recognized the mutuality of their interests and struck the 'Accord,' frustrating the managers hopes. The basis of this agreement was that no negotiations would take place without the other union being present.[15]

However, whilst at Hermes the Accord operated well, at Mercury the NUJ intervened 'in the best interests of the NGA' (NUJ FOC at Mercury). The numbers of staff to be lost from the production area, were agreed by unions and management. The NUJ and NGA agreed that the names of those wishing to apply for voluntary severance should be handed to the NGA FOC. What actually happened at Mercury, was that members were going direct to management in the hope that they would apply before the agreed quota was reached. This weakened the unions overall negotiating position and contravened the Accord, introducing a degree of tension. The NUJ FOC stepped in to prevent the problem from getting any worse. This did little lasting damage to the relationship between trade unions. However, although the unions were united on the new technology issues, significant differences still existed between them.[16] The deputy manager at Hermes pointed out the historical basis of such differences: '....traditionally printers have regarded journalists as middle class 'poofters' and journalists have regarded printers as being the 'buggers' who are interfering with what we are trying to do.'

It was felt that this conflict between the groups of employees has never subsided, but the Accord simply formally prevented conflict between the two unions, which suggests that the alliance was a very uneasy one. Problems were somewhat extenuated by the NUJ's eagerness to implement the new system, and the NGA's distaste of a new technology that threatened to take over their jobs, which encouraged them to delay implementation for as long as possible.

At both newspapers, over subscription for voluntary severance, placed the NGA FOCs in a considerable dilemma. For example, at Hermes, management agreed the loss of twenty two NGA people who would take early severance and leave the newspaper. However, many more wanted to leave, attracted by the payments being offered, and the local NGA wanted the best for its members. On the other hand, the union did not want to lose its members. Ultimately the union stood down, and around thirty nine took early severance, almost twice the original figure agreed.

Hermes Newspapers clearly had no intention of shifting responsibility downwards to the production side of the business. The thrust of the new technology changes was towards removing responsibility back into the hands of the managers. Some of the overseers were retained to take on new supervisory tasks in training and managing the inputting functions. Mercury Newspapers had similar plans. There was a tendency for the strengthening of the management and a reduction in the responsibilities of the production staff who were also used as a casual workforce. The NGA father of chapel referred to the new inputting staff as 'floating production assistants.' There was no evidence of the encroachment of lower status staff into the tasks of their superiors, and this was as true of the editorial and advertising departments, as it was of the production department.

Moreover, findings in the in chapter three, suggests that there was potential opposition amongst some editorial workers and the NUJ to the redeployment of production staff in the editorial areas, although this had taken place in other studies (Smith, 1988; IRRR, 1985).

Summary - newspaper studies

Divisions between the trade unions in the newspaper companies, were temporarily healed by the Accord for the duration of negotiations for the direct entry technology. However, within this environment, there remained a considerable degree of uneasiness because journalists wanted rapid establishment of a new system while the NGA wanted to delay deployment. Inevitably, there were internal problems within the NGA, because some of the members were concerned they would not get the early severance terms they sought. Ironically, in one newspaper it was the NGA members who weakened their own union's bargaining position, and not the NUJ as one might have expected.

Nevertheless, it is in this study that conflicts were expected to be greatest over disputes for the remaining inputting work. The possibility of an alliance to prevent inter union squabbling, which would undoubtedly have weakened the bargaining position of both unions, is an example of one means of avoiding demarcation conflicts.

Evidence is against the hypothesis in these cases, conflict being formally prevented by the Accord. However, potentially conflicts may have erupted over issues of redeploying composing workers in editorial areas. Here the possibility of conflict was minimized, but this also has to be qualified with examples in other studies where conflict did ensue, such as the Wolverhampton Express and Star dispute (NGA, undated).

The office studies provide no evidence of demarcation conflicts, although potential cases where conflicts might develop, are where supervisory responsibilities are handed down to clerical staff.

Office studies

The main possibilities, in the clerical studies, for conflict between workers and trade unions, were in the merging of work roles of supervisors with clerical workers. Evidence for this was weak, and the office studies generally provide poor comparisons with the other studies. Due to the lack of data this section uses only one general heading of demarcation lines.

Demarcation lines

The supervisory staff in the accounts department at Apollo were safe in the knowledge that the financial accountant believed in a strong management structure. He claimed that he had been in work environments that had devolved organization, and these simply had not worked. The management accountant agreed with this philosophy, but emphasized more the need for a firm management structure because of the complexity of the system. Interestingly, the payroll section was given as an example of a strong management structure in which a supervisor, deputy supervisor and two senior staff looked after a very small number of staff.

However, the payroll section had been able to organize its own work, as it was physically separated from the accounts department, and to some extent having an independent command structure. In this area, the best examples of devolved responsibility could be found. The supervisor emphasized the sharing out of the interesting and the more tedious work equally. The computer systems had successfully liberated the section from the previously very boring work that had to be done, and introduced more involving work to the section. What is important for this hypothesis, is the way that the supervisors were able to devolve responsibility in the section, but be unconcerned about any threat to their own jobs because of a management attitude that required firm supervision.

The rest of the accounts department did not reflect the situation in payroll, and in fact represented a structured division, concentrating the more detailed tasks into a few jobs, lower down the hierarchy. The computer system had also offered increased opportunities for control, and even if the supervisors did not take advantage of these mechanisms, it seemed likely that the managers would have wanted this included in the system for precisely the purpose of control.

For the manufacturing services group, a reduction in supervision on the shopfloor was not matched on the clerical side. According to the operations manager, those involved in production control had more supervision. However, it was difficult to see where there was any increase, except perhaps in the ratio of staff to supervisors, since the staff numbers had fallen and the supervisors had not. The clerical work here had increased, but more in terms of volume than responsibility. The supervisors' tasks had also increased in a similar fashion, and they had little reason to feel upset about encroachment.

The potential possibilities in Apollo to pass supervisory responsibility to clerical staff, were revealed because managers believed in strong supervision. At Achilles however, there was more evidence of vertical reskilling, involving supervisory functions, being passed to clerical workers.

The situation at Achilles was different in the distribution area, where direct supervision had been reduced, and this is a direct result of the computer system. A major part of a supervisor's job was to think about where the stock was going, and to minimize the movement of stock. These are all features that the computer systems can now comfortably handle. Supervisors had good reason to be concerned here, given the way the technology was being deployed, but little concern was evident. Most believed the system had been developed as far as it could be for the time being, and as many supervisors that could be removed had already been removed.

The accounts department, at a later stage of technological development, seemed to be on course to follow the pattern of the distribution area. Accounts' staff had been re-evaluated, and credit control staff had been placed at the top of clerical hierarchy. The credit manager confided that there would be a ceding of supervisory responsibility to the credit control staff, once the computer system was in place, and information more accessible. Supervisory staff would either lose their jobs or be redeployed.

Summary - office studies

Neither Achilles, nor Apollo revealed many problems amongst the unions in the plant. The white collar unions, APEX and MSF, stayed within their areas of, respectively, clerical staff, and supervisory and technical staff. The blue collar unions were seen as quite separate from the white collar. Only at one point, did one supervisor at Apollo suggest she would rather have one union to cover all white collar staff, and since APEX was the majority union they should represent the plant. This she maintained would give greater strength to the representation of the workforce. It was in the accounts department at Achilles where occupational mergers were likely to take place, but staff members involved, had yet to be officially informed of the change at the time of the interviews. Here, there was a possibility of conflicts between supervisory and clerical staff, and between their trade unions, respectively MSF and APEX.[17]

Little actual evidence can be found in these studies to support the hypothesis, although potentially there may be problems. Most managers were reluctant to commit themselves to job merging, particularly in the form of vertical reskilling, which made the likelihood of conflicts remote.

Summary and conclusions

This chapter, has questioned the level of influence trade unions have, on programmes for the introduction of new technology. Secondly, it has considered the potential conflicts that may occur between trade unions due to occupational mergers.

That trade unions can influence decisions in respect of new technology is not in question, it is the degree of influence that is important. There is a strong case for unions to argue, that their knowledge and cooperation in the process of change, can offer rewards both to the workforce and to the company. However, there is a tendency to view new technology in terms of its quantitative effects. Ignoring, or seeing as irrelevant, the qualitative effects on jobs (Wilkinson, 1983). This represents a narrow deterministic view on the part of the trade unions. Clearly, seeing new technology in such restricted terms, neither acknowledges the potential choice and scope managers have to change jobs, more importantly, nor does it formulate a response.

The metalworking companies, and the newspaper companies, were both examples of this tendency, but nowhere was it more prominent than at Achilles (the shoe manufacturer). Union representatives in the metalworking companies, seemed most interested in reducing the numbers of redundancies, increasing pay levels, and arguing for job allocation on the basis of seniority, rather than merit. Whilst no particular knowledge or interest was displayed in work organization issues in either company, both had attempted (and failed), to establish new technology agreements. In the newspaper companies, although there was knowledge of the wider work organization issues of new technology, representatives were mainly concerned with quantitative aspects, and did not present alternative possibilities for work organization. However, the unions of both newspapers, were stifled to a considerable degree by intransigent managers. Similarly, this was the case at Apollo (the aerospace company), where the trade union (APEX) had been active in establishing a new technology agreement, and attempting to influence work organization issues. Nevertheless, managers here saw trade unions as unhelpful, or were indifferent to the idea of consultation.

It is particularly this latter example that demonstrates that not all trade unions were found to harbour such deterministic views. Even where they were formulating effective responses, the negative views of managers acted as a barrier to discussion. Cressey (1990) was aware of this inevitable problem and pointed out that managers are the 'gatekeepers' of consultation procedures (also see Price, 1988). Therefore, it is often managerial attitudes towards trade union organization that are important, because until managers accept trade union involvement in the policy making process, no genuine consultation can take place.

In many of the cases, the importance of trade unions as potentially disruptive forces is recognized, and the response is to operate within agreements and within the minimum expected levels of consultation. This was true in Artemis (the manufacturer of precision hydraulics), Zeus (the diesel engine company), and in Apollo. But in the newspapers, managers knowledge that they now had the upper hand, meant that a narrow 'quantitative agenda' was set for negotiation. In the other companies, what managers called 'consultation' was actually no more than information provision, often used instrumentally to prepare workers for future change. Of all the cases, trade union organization proved to be least effective at Achilles, managers had taken advantage of the lack of knowledge and narrow view of the union representatives.

Trade union involvement depends both on the ability of the trade union to make a case for inclusion in discussions (Price, 1988), and on managerial acceptance of trade unions (see chapter two). In the case of Apollo, trade union 'sophistication' (Batstone *et. al.*, 1987) is well developed, but managers' refusal to accept union involvement prevents constructive debate. This runs counter to the hypothesis, which maintained that unions would be guilty of failing to make an effective case for inclusion in discussions. Whilst much of the evidence agrees with the hypothesis, it has to be accepted with the reservation that lack of consultation depends, less on effective trade union organization, and more on the attitude of managers to union inclusion.

Part of the analysis in chapter five, considered the nature of jobs following new technology introduction following theoretical analysis in chapter two. This established, that some of the main job changes, were represented by additions of new tasks at equal, or higher levels, than the original job. However, the consequences of this occupational merging inevitably means the removal of skills from one worker's job, and allocation to another. In this relationship, conflicts are likely between both the workers themselves, and between their trade unions. The second hypothesis, presented in this chapter, claims that this is an increasing occurrence.

There is an important link between trade union responses to new technology change, and their response to this particular problem of demarcation conflicts. Other commentators have also referred to this issue (for example Jones, 1988; Rainbird, 1988). However, some versions have encountered the problem in one particular sector (for example Jones, 1982 and 1983), while elsewhere, the focus has been more from the point of view of national union organization (Rainbird, 1988). The hypothesis offers an opportunity to consider the incidence of demarcation conflicts across the range of cases and at a local level.

Various occupational mergers were occurring in the companies. Some of the best examples were to be found in the metalworking cases, where operators and programmers' jobs (see also Jones, 1983; Wilson and Buchanan, 1988; and chapter three), and maintenance workers' jobs were merging. Conflicts had developed at Zeus over the latter, in the former there were clear signs of potential conflict. However, at Artemis programmers felt confident that job merging would extend no further than the addition of amending tasks to operators' jobs, and this prevented conflicts. It was at the newspaper studies where conflicts seemed most likely, given the difficult history between NUJ and NGA, and the proposal to divert inputting tasks away from compositors to journalists. However, here an alliance developed to avoid just such conflicts. In the office studies, evidence of demarcation breaching was rare, and evidence of union conflicts was non existent. Only Achilles demonstrated any possibility of union difficulties, as a result of high grade clerical staff taking over supervisory functions. Here APEX represented the clerical workers, and MSF the supervisors. However, there was no evidence of any knowledge in the trade unions of the prospective changes, quite apart from any conflict over the incident.

Evidence for conflicts arising from occupation mergers is not strong in these studies, although there are a number of cases where demarcation lines have been breached, and conflicts remain a possibility. The hypothesis can be accepted only reservedly, but there are clearly important implications for trade unions. The NGA and NUJ discovered the damaging nature of such conflicts in disputes early in the provincial press's new technology campaign (NGA, undated; IRRR, 1985) and responded with the forming of alliances. The effect of this, was to undermine managerial confidence in using the rift between the two unions, to divert attention away from the changes. Hence, despite a lack of evidence of actual conflicts, potential conflict situations are created if demarcation lines are breached. Trade unions must be prepared to deal with this. Since conflicts are possible, even within one union, this implies that even more sophisticated responses may be required than alliances of the kind in the regional press.

Notes

1. Enterprise unionism refers to a management style system of representation unique to the particular enterprise in which it is organized. Workers become members and elect representatives who consult with managers but there is no affiliation of the organization to any trade union or the TUC. The concept is derived from Japan. See Brown (1983).
    NB. Opposition of quality circles, another Japanese import, were being expressed as AEU policy at the time of the interviews.
2. The process of transferring from hot metal to production take photocomposition and then to a full direct entry system is described in chapter four.
3. Chapter four gives a much fuller explanation of the jobs and the changes.
4. Evidence suggests that this was not a voluntary early retirement but was enforced by the parent company Westminster Press. See chapter six.
5. This was largely because the many errors in advertisements were damaging for reputation and cost dearly in credit notes.
6. An agreement at Apollo had established 'across the board' increases for any work involving VDU operation.
7. Compare with the Zeus case study.
8. See Rainbird (1988) and chapter two.
9. Other examples of this might be the need to use programmers with the introduction of CNC, so that some of the work of the traditional craftsmen is now embodied in a program written by somebody outside the original task structure. Other examples are the blurring of the electrical and mechanical maintenance functions at Zeus and the direct entry of text by journalists in newspapers. In the clerical studies the restructuring of staff to displace supervisory roles is also an example.
10. These were other skilled people such as plumbers, for example.
11. SOGAT '82 have since amalgamated with the NGA to become the Graphical Paper and Media Union.
12. Direct entry changes refer to the change in new technology that would allow the editorial staff to directly input 'copy' into a computer system instead of going through compositors.
13. The pre-entry closed shop meant that not only did all skilled production workers have to be a member of the NGA but they had to be on the books of the NGA before they could be employed. The employers would not advertise on the open market for a suitable person but would ask the NGA, who would find somebody from their register.

14. One of the best known disputes was at the Wolverhampton Express and Star, see The Wolverhampton Express and Star: a briefing paper (NGA, undated).
15. Fuller details about the 'Accord' are provided in chapter four.
16. For example the NUJ FOC referred to the traditionally sexist attitude that the NGA harboured, and which has been studied by Cockburn (1983).
17. Unfortunately, because this information was offered in confidence (and with the tape recorder turned off) I was unable to take up this development with the supervisors and staff directly, although from some of the comments it seemed likely that a change along these lines would not come completely unexpected.

8 Summary and conclusions

Introduction

This book set out to cast further light on the effects of new technology on work organization. The specific aim has been to address issues that influence and inform the decisions for changing jobs, after the introduction of new technology. Hence in general terms it looks at the work organization aspects of managing new technological change, but more specifically attempts to identify, and solve, some of the issues and problems of redesigning jobs.

Chapter one identified a key problem as the existence of choice which prompted dilemmas and debates for work organization issues. This study has considered some of these choices from the choice of new technology, choice between styles of work organization, and choice in the allocation of workers to jobs. These choices are variously influenced by actors and groups, and the conduct of procedure in the approach to change. The former include trade unions, and the latter the planning that goes into change.

The study of this area itself is informed by controversies and gaps in the existing theoretical literature and in case studies. Findings both confirm and challenge existing ideas, and in some areas, extend the existing knowledge of the field.

Background to the study

Following a discussion of the issues and objectives, in chapter one, chapter two took up theoretical issues in the existing literature that dealt with methods of job design, new technology and its effect on skills, and finally industrial relations aspects. This examination revealed a number of areas of contention and formed a foundation for the study. Chapter two was followed by an analysis of existing case studies in chapter three. Here the three areas of managerial intentions, work organization, and trade union involvement in technological change, were used as a structure for the analysis of empirical findings. This was conducted across the three new technology categories selected for later investigation, with the use of new case study evidence. Finally, a section at the end of the chapter postulated five hypotheses based on the deficiencies in literature and in case study reports.

These hypotheses were not tackled immediately but one chapter (chapter four) was first devoted to a review of the companies selected for analysis. This includes the background details of the firms, the new technologies and the jobs. The purpose of this chapter was to provide information in a broader context prior to presentation of the data, to assist the readers' understanding of the subject organizations.

The leading hypothesis chapter (chapter five) dealt with two hypotheses. The first of these concerned the nature of work in the case study companies following the introduction of new technology. Following claims that new technology promotes particular work design forms (Braverman 1974; Hirschhorn, 1984), this section considered the changes in the companies suggesting that the most likely scenario is a mixture of skill outcomes. In the second part of the chapter, the hypothesis claimed that the allocation of work following the introduction of new technology, was substantially influenced by social and political criteria aside from technical criteria.

One main hypothesis was broken down into three sub hypotheses in chapter six. The extent to which the technological change was planned was the theme of the first part, focussing particularly on work organization issues. The second section draws on the first, claiming a lack of strategic planning results in a series of *ad hoc* approaches which in turn provoke a range of rationales not necessarily conforming with the organizational aims. Finally, the chapter turned to consider levels of involvement as an essential ingredient in this planning process. Here the hypothesis maintained that decisions concerning new technology issues generally, and specifically the work organization aspects, were confined to senior managers and little consultation was held with the workers and prospective users.

The final hypothesis chapter (chapter seven) pursues a further two hypotheses both dealing with trade union issues. Following on from chapter six, the first of these claims that extent of trade union involvement in new technology decision making would be limited. The second hypothesis suggested that programmes for the introduction of new technology may mean the merging of jobs that breach occupational demarcations which will in turn lead to conflicts between and within trade unions.

The nature of work

Hypothesis - The diversity of factors governing the organization of work with new technology means that neither deskilled or reskilled work will predominate even in the same firms or industries.

Findings in chapter six and existing work in the field (Child, 1972; Rose and Jones, 1985) suggested that strategy is incomplete as far as the definition of work organization and task structures are concerned. It seems unlikely that case study companies will be following a specific job design ideal. Job design implies planning, logically if there is no planning there is no design style. However, the argument does not maintain that planning is completely absent. Elements of planning may inform ideas about the best form of job definition. Secondly, to adopt a technologically deterministic approach and argue that the technology defines the jobs will eventually result in managers making judgements about what the new technology dictates. The work organization that emerges may then be comparable to models presented in chapter two.

The hypothesis maintains that there will be no predomination of deskilled or reskilled work. If confirmed, this would seem to rule out Taylorism and the predictions of many labour process theories since these both refer to deskilling as a central feature. Other studies confirm this finding (Jones, 1982; Batstone *et. al.*, 1987; Wainwright and Francis, 1984). However, it is worth noting Armstrong's (1988) qualifications of Braverman's (1974) work where he suggests deskilling is a general trend that may not be detected by a short term analysis. Therefore, accounts appearing to refute Braverman's hypothesis (Jones, 1982; Burnes, 1989; Wainwright and Francis, 1986) are not acceptable. Lack of strategic planning also denies the existence of sociotechnical approaches, since the nature of this style is designing jobs with both technical and social criteria in mind and requires considerable planning. Study findings suggest that social criteria were not considered alongside the technical aspects, which were, in many cases, virtually an exclusive consideration.

In the case studies there was overwhelming evidence of horizontal reskilling (the addition of new skills of equal complexity to workers current jobs), present in all the companies and across a broad range of jobs. There was little solid evidence of vertical reskilling (the addition of skills at a higher level than the worker has been accustomed to), but there were hints of this in the operator programming of CNC machines. Consistent with work by Hirschhorn (1984) and Touraine (1962 in Littler, 1982), clerical staff were taking supervisory responsibilities. Elsewhere a possibility that never materialized was in the newspaper companies where composing staff made a bid for editorial jobs. There was some evidence of deskilling and this was mainly in the newspaper and office cases.

The findings also suggest that it is possible for deskilling and reskilling to be found in the same company and in the same department. This could imply the emergence of a division between primary and secondary workers (Edwards, 1978). In the newspapers and office studies the workers involved had different employment contracts and could therefore be categorized in the Atkinson (1985) model of numerically and functionally flexible workers.

Alongside these developments was the possibility that the findings are consistent with predictions of a post Fordist approach, to apparently enhance work while actually increasing levels of control over workers (Palloix, 1976; Aglietta, 1979; Pignon and Querzola, 1976; Ramsay, 1985). Therefore, cases of horizontal reskilling may be accompanied by motives of control, and in many cases there is evidence of this. For example, in the accounts departments of Apollo (the aerospace company) and Achilles (the shoe company), where managers reduced personal contact, by making the information more immediately available, and by using the system to gauge staff performance. Also in the metalworking companies, where self inspection increased accountability alongside the task content of the workers.

The recent reassertion of the Braverman thesis (Armstrong, 1988) contrasts sharply with claims that the future of work will involve more, rather than fewer skills (Cavestro, 1989; Kelley, 1989). The evidence here tends to agree with the latter rather than the former. However, evidence of vertical reskilling, also referred to as 'flexible specialization' (Piore and Sabel, 1984), and 'worker centred control' (Kelley, 1989), in the case studies is rare. Where there appears to be evidence, closer analysis suggests that managers fail to cede full authority, this is possibly because they are concerned about the loss of control involved in such a process.

Horizontal reskilling is more easily identifiable. This is what Kelley (1989) refers to as 'shared control.' Shared control involves some aspects of planning being taken by workers. The case study job changes might otherwise be categorized as 'polyvalence' conforming to the Jones and Rose (1986) and Child (1985) definition in which workers are able to perform a range of tasks that cut across traditional skill and job boundaries, but not necessarily implying the addition of higher level skills.

Examples exist in all companies of new skills being added to jobs. However, in several cases, particularly the newspaper and office studies, specialized work existed alongside the polyvalent workers. In both newspapers and in Achilles, workers were used on part time contracts recalling the Atkinson (1985) model of functionally and numerically flexible workers (see also Atkinson and Meager, 1986). Another important feature of the numerically flexible group, particularly in the newspapers, was the overwhelming gender orientation. Women are seen as a flexible and cheap workforce and therefore attractive to employ (see Walby, 1989).

The hypothesis about job nature started out by claiming that there would be several forms of work organization. This hypothesis is confirmed but there is clearly a dominant form of work organization in the case studies of horizontal reskilling. Alongside several examples of reskilling suggesting polyvalence are hints of a post Fordist structure. Here aspects of new technology may be used to enhance jobs, but at the same time employ methods of surveillance, to assert greater control (Jones and Rose, 1986; Pignon and Querzola, 1976).

Work allocation

Hypothesis - Allocation of work is not always based on criteria of finding the most technically suitable person for the job but political and social criteria are also involved.

That new technology and job design are socio-political decisions, as well as technical and economic ones, was established in chapter two (see Wilkinson, 1983). The allocation of work is also subject to similar processes. Instead of being a consideration of the technical qualifications of staff alone, managers often take into account other non technical criteria.

The hypothesis discussed in chapter five considered these criteria under the headings of 'social and political' and 'technical' criteria. In the case of the latter, criteria was manifest in technical qualifications, experience and 'ability.' For example, Artemis (the manufacturer of precision hydraulics) called on high qualifications in its new employees. The newspapers were concerned about compositors' ability in terms of skills, to transfer to editorial work. The office studies expressed preferences to retain existing staff, rather than employ new workers, because of the experience they could offer. Nevertheless, in each company reviewed, either social criteria, political criteria, or both of these, were applied alongside the technical criteria.

Age appeared as a deciding factor in the four manufacturing companies, where managers suggested that older people were not capable of assimilating new technology. Not only were older workers generally not selected for work with new technology but they were also earmarked for redundancy. However, there was evidence to suggest that older workers were capable of handling new technology and retraining, since they were handling VDU work in Apollo (the aerospace company), and at Zeus (the diesel engine manufacturer) were part of the retraining programme interchanging maintenance skills.

The reasons for rejecting workers on grounds of age could also be seen as technically motivated, since managers were complaining that training older workers was wasted or at least offered a significantly lower return on their investment than training younger workers. However, it seems likely that younger workers with up to date training would be attractive to other companies and inclined to leave the companies for which they worked far more quickly than older workers. One example of this was the maintenance craftsmen at Zeus who, because of their broader repertoire of skills, were becoming attractive to other companies. Jones (1988) has also identified the possibility of this happening.

Trade unions are also capable of influencing the allocation policy of an organization, and in some of the companies were pursuing management acceptance to their views. One example was the AEU at Zeus who favoured a system of promoting the older, more experienced workers. However, at the same time the trade union wanted to protect staff from embarrassment if workers felt they could not handle training. This appeared to be rather contradictory. Whilst prepared to accept the trade unions 'no embarrassment clause' managers were not prepared to consider allocation on the basis of seniority. They nevertheless, applied social and political criteria of their own, reflecting biases against older workers.

Other examples of where trade unions have been able to influence the allocation of workers was at Hermes and Mercury (the newspaper companies). Here male compositors had monopolized skilled production work as in Cockburn's study (1983; see also Phillips, 1983). NGA compositors had no experience of standard typewriter keyboards that were introduced with photocomposition systems. Here the most appropriate allocation of work would be to employ new staff with secretarial skills whose inputting abilities were far more efficient. This would have probably have meant employing many more women. The trade union had effective influence on allocation policy preventing employment of what seemed to be the most appropriate group of workers.

More advanced forms of new technology meant that NGA workers had to relinquish their monopoly on composing work. At this stage the allocation criteria became socially influenced by managers. Women were welcomed into the companies to handle inputting tasks, they were on short term contracts with a fraction of the pay of the NGA workers. This view of women as a cheap casual labour force was also prevalent in both of the office studies. In Achilles women were used as a 'stop gap' during a transition to new technology (see Walby, 1988).

Politics was evident in allocation criteria where companies made value judgements based on the attitude of individuals towards the company. This occurred in both of the metalworking companies and in the two office study companies. In the newspapers the political exclusion was acute, and based largely on a historical distaste of the NGA. When 'direct entry' new technology was introduced making compositors jobs superfluous, compositors were prevented access to editorial areas due to strong biases amongst journalists and fear of disputes over the levels of pay that compositors would be receiving (which were linked for the first year to their previous jobs). Technical reasons were offered for their rejections, but the fact that similar redeployments have been successful elsewhere (Smith, 1988; IRRR, 1985) leaves this in some doubt.

Far from being the most technically appropriate allocations, evidence suggests that a myriad of social and political factors are also involved in work allocation decisions. However, it did not seem to be the case that a high standing in technical qualifications exempted workers from any part of the social consideration, rather the social consideration was capable of overriding technical criteria. Hence the allocation hypothesis is confirmed, even to the extent that in some cases, the predominant criteria for work allocation may be social and political, rather than technical or economic.

Strategy, rationales and participation

Hypothesis - In the work organization aspects of programmes for the introduction of new technology, formally planned strategies for change are rare. More often changes are dealt with on an *ad hoc* basis as and when problems arise. Rationales for the use of new technology are varied and do not reflect any general motive of managerial control over the workforce.

There is an inextricable link between planning and the actors involved in this process. The latter part of this hypothesis maintains that any consultation involving the planning of work organization that does take place is confined to senior management levels and rarely reaches the workers and users.

Strategy

Findings from studies such as Rose and Jones (1984) question the pervasiveness of planning. Consistent with this, technological determinism, which implies a lack of planning because it is believed that the technology is the determining characteristic, was prominent as a feature of management opinion in many of the companies. The belief was that as long as the new technology was carefully considered then there was no need to separately plan job design. However, Buchanan (1986) points out that technology is not a determining characteristic but an enabling one. This is elaborated by Buchanan and Boddy: 'It is more accurate and appropriate to view new technology as a *trigger* to the process of organizational decision making in which critical *strategic choices* emerge' (Buchanan and Boddy, 1986; emphasis in original). New technology creates certain opportunities and may also create constraints, but it is within a wide range of options that decision making is fluid (Child, 1972).

Most companies failed to recognize this and adopted a technologically deterministic view. Both the newspaper and office studies adopted the view that new technology would make the work organization decisions for them. In these studies, and in the metalworking studies, there was considerable planning for the systems and machinery. However, lack of planning for jobs at Artemis (the precision hydraulics company) linked to the new technology developments, was based on the assumption that since workers were already operating new technology no planning was required. This overlooked the greater complexity of the new machinery and job allocation, and the fact that new people were required, in addition to those already operating machines. Zeus (the diesel engine company), had a rather more instrumental purpose for job planning, to add new tasks to jobs in advance to avoid the inevitable future pay claims for the performance of extra functions involved in new technology usage.

In most cases the importance of job planning is either underestimated as in Artemis, ignored as in the newspaper companies, or misdirected as could be argued in Zeus. It is naive to think that something as potentially complex and flexible as new technology will only offer one set of options for work organization (Buchanan and Boddy, 1986) but equally a problem is the limited nature of planning. However, given that these opinions persist amongst managers, work organization is likely to be subject to a series of incremental changes as occurred in one department at Apollo (the aerospace company).

Rationales

Related to how the new technology is introduced are questions of why it is introduced. Often there are aims defined at corporate levels of an organization. These aims are sought by means of strategy. However, strategic planning for job changes is often limited or non existent, and this leaves managers in specific departments to decide on the definition and organization of work. As a result of the autonomy of managers actions, the justifications that inform their decisions, are not necessarily related to the overall aims. In some cases the justifications of managers may subvert the organizational goals if managers wish to pursue their own ideals (Buchanan, 1986).

There was not much evidence of the latter, but in some cases, such as the accounts department at Apollo, the management seemed to be justifying new technology on the basis of control which had apparently caused embarrassment for the personnel department. Elsewhere, in the newspapers and metalworking studies, control and accountability were referred to as rationales. In the newspapers, the organizational aims seemed to be conducive with such motives, but this was not necessarily the case in the metalworking cases. On the whole, the diversity of rationales suggested that lack of strategic planning, had resulted in the need for a series of retrospective justifications.

There are two points here. The first is that lack of strategy creates the environment for diversified *ad hoc* decision taking resulting in the definition of a new set of rationales peculiar to a particular department and not necessarily consistent with organizational aims. The second point is that even within a framework of strategic planning managers may pursue goals that reflect personal preferences which may diverge from the strategic aims defined at corporate level.

The problems are similar in nature to those identified in the previous section. Neglect of planning necessitates an incremental approach by managers. The incremental measures are based on managerial perceptions and may not result in work organization and effective job definitions. Alternatively, managers may pursue objectives that actually run counter to organizational aims but are of special interest to the manager. Hence, in the case studies, there are examples of managers employing rationales of control, while there is no evidence of corporate motives that necessarily require such an approach. This according to Buchanan (1986 p75) is 'dysfunctional in terms of operator skill, motivation and performance.'

Involvement

Questions of how and why new technology is introduced would be incomplete without questions of who is involved in the decision making process. There are many possible actors including parent companies, senior managers, line managers, employees, employers associations, and trade unions. The key importance of trade unions warrants closer examination and a separate hypothesis was reserved to deal with this issue.

The influence of these parties can alter the organizational aims and change the course of strategic approaches. Workers and trade union participation is often acceptable at the level of implementation but rarely at the level of devising management strategy (Cressey, 1990; Northcott *et. al.*, 1985; Cressey *et. al*, 1988). Indeed, Jones (1988) has confirmed that workers are rarely involved in discussions about the new technology to be introduced. Yet their inclusion is essential because their 'on the job' knowledge may include important unarticulated details of their work that management are not aware of (Libetta, 1988).

In some of the companies managers had recognized this, and had developed direct communication with workers in preference to communication with trade unions. This was taking place at in Zeus (the diesel engine manufacturer), Artemis (the precision hydraulics company) and in the office studies where information about prospective changes was passed direct to members of staff. Zeus suggested that this style undermined the need for the trade union stewards to disseminate information and was therefore a reassertion of managerial control. Rainbird (1988) has further suggested that quality circle arrangements are a means of tapping the tacit knowledge of workers. The AEU at Zeus felt threatened on this basis and management union relations had suffered as a result.

Other intervention from parent companies occurred at Zeus and Hermes (the smaller newspaper). The implications of such an approach were that the choices available to the companies were restricted by the demands of the parent. In the case of Zeus this was a warning to improve profitability which undoubtedly acted as a catalyst for both the introduction of new technologies and reforms in work organization. Hermes experienced intervention that changed the direction of managerial policy towards negotiations. Elsewhere, workers' involvement in situations where companies claimed that the new technology determined the job definitions seemed irrelevant. In Apollo (the aerospace company), one manager suggested the lack of knowledge of workers rendered them incapable of making any constructive contribution.

Generally however, confirming the hypothesis, new technology discussions were confined to higher level management, although supervisory staff were sometimes consulted about worker knowledge. The repercussions for this kind of approach are likely to result in uninformed decisions about the reorganization of work. Combinations of limited planning and poor staff consultation suggests that work organization is likely to result in a lack of fit of workers to jobs, this in turn may lead to inefficient and ineffective work organization.

Trade union participation

Hypothesis - Trade unions primarily associate new technology with quantitative issues such as job loss, severance pay and new pay levels. Lack of knowledge and resources leads to ineptitude for making a case for trade union inclusion in discussions of the qualitative effects (work organization, job design) of new technology.

Trade unions, as representatives of the majority of workers in the case study companies, are potentially an important force in diverting the direction of change. Given that the choices and decisions made by managers are highly fluid in nature then it is not unreasonable to assume that a trade union may be influential in the change process. Cressey *et. al.* (1988; also Cressey, 1990) have argued that there are a number of possibilities for participation in the process of introducing new technology.

Broadly these possibilities range from no involvement and notification only, through to negotiation, consultation, and ultimately co-determination.

There are several factors emerging in the case studies that suggest there are particular elements influencing the extent of involvement. One aspect is what Batstone *et. al.* (1987) describe as 'union sophistication.' This means, particularly, that trade unions have to see that they themselves have a place in changing and influencing the choice of new technology system, and the organization of work. Many are too poorly resourced to handle this (Jones and Rose, 1985), while others fail to see their objectives extending any further than the quantitative and tangible aspects of new technology change and issues of pay increases, redundancy levels and severance payments (Northcott *et. al.*, 1985). Trade union preoccupation with quantitative issues was evident in all of the companies with the exception of Apollo. However, at Zeus and Artemis, the AEU had attempted to establish a new technology agreement (NTA) but managers had refused this. At Artemis this was because managers feared they would be restrained by such an agreement.

Another factor that directly effected the extent of trade union involvement was managerial attitudes to new technology. At Zeus, managers eventually agreed to a NTA but this was conditional on the union's acceptance of the use of sub contract labour in the plant, and the AEU rejected this condition. Apollo had already established a new technology agreement but little of this was adhered to. Apollo, like Zeus, and the newspapers suffered from adversarial managerial attitudes towards trade unions. Apollo could see little use in including the union in talks, while Zeus tended to use a consultation style which took the form of advising workers about changes but not listening to their views. These findings confirm work by Williams and Steward (1985) on NTAs, showing the limited application and effect of such agreements.

Managers in the newspaper companies, also viewed trade union involvement negatively. The agenda that had been set for negotiations included issues of redundancy levels, redeployment terms and severance pay. Neither managers nor the trade unions introduced issues of work organization. At Apollo, the difference was that the trade union had considered work organization issues but managers prevented them from having any involvement. Achilles stands out as a company with an intransigent management, but also with an ill informed and apathetic union.

Trade unions generally failed to put forward strong cases for inclusion in discussions of work organization issues. There was a lack of knowledge of these aspects and union representatives tended to identify with quantitative concerns. However, management attitudes to the trade unions had played an important role. Even where the union is capable of making a strong case, management's approach can negate this. As Cressey (1990) points out, managers are the initiators and the gatekeepers of such processes and control what happens, even if there is a strong trade union case for involvement. The solution to this problem for trade unions is by no means a simple one at organizational level, but managers might be won over by the union's presentation of workable alternatives to managerial policy.

Price (1988) suggests that much greater resources are required if trade unions are to achieve coordination and respond proactively (see also Jones and Rose, 1985). The hypothesis claiming the inadequacy of trade unions is accepted reservedly, since even given knowledge and resources required, there is not automatic inclusion in new technology discussions without managerial acceptance of a trade union role.

Occupational mergers and traditional demarcations

Hypothesis - As new technology is introduced, occupational mergers take place which may mean the breaching of traditional demarcations. Where this happens conflicts will develop within and between trade unions.

As evidence above has shown, technology changes jobs. In some cases this means a merging of jobs or tasks which can cross traditional job demarcation lines. The consequences of this approach may be conflicts between groups of staff and their trade unions. This could mean the blurring of boundaries between direct and indirect workers as noted by Arthurs (1985) and Coriat (1988). For example, in studies by Wilson and Buchanan (1988) and Jones (1982; also Jones, 1983) merging of programming tasks with CNC operators jobs, resulted in disagreements between the programmers' trade union (MSF) and the operators' union (AEU).

Like these studies, Artemis and Zeus experienced occupational mergers involving the transfer of programmers skills to operators. However, production engineers (the existing programmers), only felt threatened in Zeus. This was because Artemis had failed to commit itself to full operator programming, despite claims by some managers that the ultimate goal was to bring operators to a level equal with production engineers in terms of their skills. In Zeus the existing programmers were going to great pains to demonstrate their distaste of operator programming that went beyond simple amending work. This had not yet developed into a dispute between the MSF and AEU.

Other possibilities of occupational mergers are likely where maintenance workers exchange skills (Jones, 1988). This occurred between electrical and mechanical maintenance workers at Zeus. The EETPU objected to mechanical maintenance staff learning their skills. Without success, they tried to impose restrictions on the extent of the electrical work that multi-skilled workers could do. Also at Zeus occupational merging, of 'skilled' maintenance tasks with 'semi-skilled' operators jobs, was rumoured. Both the AEU and EETPU were angered by this possible merger of skilled work with semi-skilled jobs. Since the semi-skilled were also members of the AEU, this provided the possibility of internal union conflicts.

It was noted in chapter five that,at both Zeus and Achilles (the shoe company), supervisory authority was being handed down to lower levels of worker, made possible by the greater ability of managers to use new technology for surveillance purposes (Jones and Rose, 1986). This was one of the few examples of demarcation changes in the office studies, but neither here, nor in Zeus, did it seem to be inciting either trade union or worker conflicts.

Chapter four considered the adversarial relations that had existed between the NGA and NUJ in the newspaper industry as a background to the study. Some examples of such conflicts are recorded by the NGA (undated) and the Industrial Relations Review and Report (1985). Based on this history, newspapers were expected to display conflicts between the two major unions and NGA, over the merging of part of the composing function with the work of journalists. In fact the opposite happened and a temporary alliance between the two unions prevented conflict.

The hypothesis maintained that conflicts will develop as a result of occupational mergers that breach traditional demarcations. Confirmation of this claim has to be qualified. While there were cases of potential conflicts developing, actual conflicts were limited. Moreover, in the newspaper studies, instead of a conflict situation an alliance developed. Rainbird (1988) suggests that this is the kind of approach that is required to deal with the shifting job boundaries and maintain a cohesive organization, although she seems to be referring to a more permanent amalgamation of trade unions. However, demarcation disputes that occur within trade unions may not be so easily resolvable.

Conclusions

This project began by acknowledging that new technology changes jobs. Rejecting technological and economic determinism, chapters one and two argued a range of choices exist in the technology employed and in the organization of work (see for example Davis and Taylor, 1976, Littler and Salaman, 1984; Wilkinson, 1983). This was the conceptual basis for case study analysis which looked at the management of new technological change with specific reference to job design and work organization issues.

Many managers, in the case study companies, themselves adopted technologically deterministic attitudes. As the hypothesis in chapter six maintained, planning work organization was significantly under played. Once the technology itself was chosen little additional planning was deemed to be necessary. The idea that particular forms of new technology dictate different design styles, echoes the deterministic approaches of Woodward (1965) and Blauner (1964).

From a rather vague series of strategic aims identified in the companies a series of rather more specific and more numerous rationales emerge. Why this should be so is precisely because the questions of work organization are laden with choice. The corporate organization may have failed to define aims in any concrete or specific manner, but if this is the case the decisions have to be made somewhere. Here the decision making responsibility is pushed down the organizational hierarchy and broad corporate aims are reinterpreted by line managers, and in some cases ideas are pursued that diverge from the intended corporate aims. There are exceptions to this in cases where, because of their knowledge and expertise, line management in production areas were involved in technical decisions as of necessity. This is why there is sometimes disparity between what corporate departments and directors see as the organizational aims and rationales, as compared to those of line managers. The problem is a general failure to acknowledge that there is choice in work definition which ultimately results in a series of *ad hoc* responses by line managers.

Similarly, it follows that if new technology defines jobs there is no need to consider the views of others. This means that key actors, principally the users of new systems, are not consulted. In some of the case study companies there were direct forms of communication such as quality circle arrangements, often these were used for the presentation of management information rather than the exchange of views. Managers were unwilling or unable to recognize that there were elements of knowledge that only workers were party to, which may be crucial in understanding jobs, and needed to be taken into account when redesigning work. In several examples managers' beliefs that there was no choice in work organization led to the conclusion that discussions with staff were pointless. Inevitably, this had meant, and was likely to mean, a series of incremental changes to work organization and jobs.

In chapter two a number of job design theories were presented. Amongst these Taylorism, Fordism as management perspectives and the labour process interpretation stand out as being models in which job definition involves no dilemmas and as such is unproblematic. In their own ways each of these ideas are deterministic, seeing management practice as having only one option. Sociotechnical design theorists (Davis and Taylor, 1976) later challenged this determinism and some labour process writers have modified their idea that Taylorism is the exclusive management method, and now refer to neo Fordism and post Fordism as management methods capable of winning control over workers in a more subtle manner (Palliox, 1976; Ramsay, 1985; Aglietta, 1979).

Managers' use of control systems provides an example of one choice, but others may choose alternative options. In chapter five evidence across the six case studies, examined in an hypothesis about the nature of work, suggested that the broadening of jobs with the addition of new skills (horizontal reskilling) was common. While significant deskilling took place in the newspaper firms, reskilling was also occurring. Elsewhere, notably in the two metalworking companies, and Achilles (the shoe manufacturer), there were examples of increases in the status of skill as well as the quantitative increase (vertical reskilling), although these were not particularly strong examples. The point is that in each of these cases a range of choices are available and these and other studies clearly demonstrate the range of alternative options as the hypothesis maintained they would. It is therefore misleading to argue from the position of one management model or sociological interpretation since the evidence suggests that there is a considerable variation of options. Furthermore, even with such variation of job definition, managers themselves were claiming that technology was determining work organization. However, consideration of work allocation criteria revealed that managers and trade unions were exercising choice, albeit in an *ad hoc* manner. Work allocation was the subject of another hypothesis and requires further elaboration.

Based on technologically deterministic ideas, workers would be allocated in strict accordance within the technical demands of this system. While managers in the studies often maintained that the deployment of personnel was based on technical criteria, the evidence contradicted this. In chapter five there were clear demonstrations of the intervention of both managers and trade unions to achieve allocations of workers on the grounds of both social and political criteria, in accordance with the hypothesis. Here a range of choices are exercised freely, and in several cases the apparently more 'suitable' workers are rejected on political or social grounds in preference for others. In fact in the six studies the non technical criteria became so important as to shadow the technical criteria.

Trade unions often attempted to intervene in job allocation issues. But like managers, union representatives viewed new technology in a very deterministic manner. New technology was seen, in many cases, as being responsible for job loss, and in several cases trade unions focused on this aspect and attempted to formulate responses accordingly. This involved minimizing the loss of workers on the one hand, and maximizing redundancy payments on the other. This was consistent with the hypothesis and other trade unions had also become more involved in the quantitative issues, demanding pay increases for using new technology, and Zeus (the diesel engine company) had itself devised a response to deal with this. However, a majority of trade union representatives in the studies failed to see the possible choices available in work organization, believing new technology would shape work design. Hence, in most examples there was no case put forward for alternative forms of reorganizing work. Union representatives at both of the metalworking companies even expressed preferences for highly detailed work that required little responsibility.

However, the AEU at both the metalworking companies had made failed attempts at establishing new technology agreements, demonstrating that there were obviously moves to establish more involvement in new technology introductions. Such an agreement had been established at Apollo (the aerospace company) but was not adhered to by management. Therefore, in some cases managers were just as unreceptive in cases where trade unions did recognize alternative possibilities in work organization. In order to innovate, both parties clearly have to recognize the choices that are available while introducing new technology. Choice is also available in the extent of the trade union involvement (Cressey, 1990) but until now managers have chosen low levels of participation. It is managers who hold the key to including trade unions, but trade unions themselves have an important role to convince doubting managers of the advantages of consultation and co-determination. Just as managers have tended to shy away from consultation, so trade unions in the majority of the case studies have been ineffective at formulating a strategy to argue for involvement.

The dangers of trade union preoccupation with the quantitative issues of new technology, are well illustrated by the various potential conflicts in the companies, and actual conflicts occurring between trade unions that emerge due to occupational mergers that involve demarcation breaching, an argument of the hypothesis in chapter seven. Trade unions were clearly poorly prepared for this although the NGA and NUJ had responded by forming an alliance following damaging conflict in the newspaper firms. If the incidence of this increases, there is a need for trade unions to acknowledge that technology is blurring job boundaries and they must respond, possibly by examining other options in work organization. Another option is the forming of alliances to prevent those rifts between unions which cause vulnerability, lessening the incentives for managers to consult with trade unions (TUC, 1979; Rainbird, 1988).

Technological determinism has been superseded as a means of understanding the way work is organized following new technology introduction. However, evidence shows that it is still employed by managers and trade unions. Yet they both actively pursue choice in work situations, intervening in what would be the 'pure technical solution.' This seems rather contradictory, but it stems from a realization that new technology is not capable of automatic job definition at all. In fact lack of prior planning necessitates a series of post introduction, incremental changes at different levels of an organization reflecting social and political preference, as well as the most 'technically suitable' solutions. This book highlights the key importance of planning, for managers and for trade unions at local level. This also involves an acceptance that real choice exists for variation of work organization, otherwise there is nothing to plan for. Extending the argument further, it is essential for managers to call on trade unions and workers as groups capable of highlighting alternative choices. This is as important for managers as it is for workers as several examples from the study areas illustrate. Batstone *et. al.* (1987) in their study of CNC, Cockburn (1983) and Smith and Quinlan (1982) in their studies of the Croydon Advertiser and Wainwright and Francis (1986) in their example of the education college study, all demonstrate the beneficial effects of co-determination on labour relations. Developments in UK firms' adoption of quality circles provides evidence of the realization of worker innovative ability. However, in the Zeus study the quality circle arrangement was employed by managers to communicate with workers and bypass the trade union. Elsewhere (Rainbird, 1988) quality circles have been seen as means of tapping workers tacit knowledge. Sabel's (1982) assertion applies in these cases, he suggests that managers are attempting to gain a high trust response from a low trust situation (high and low trust designation from Fox, 1974).

Nevertheless, from studies where significant levels of genuine co-determination have taken place it is possible to conclude that management/worker relations benefit from such 'high trust' relationships. These forms of cooperation encourage worker innovation because workers recognize that they derive benefits directly from the ideas they present.

Planning may not be the panacea to solve work organization problems in managing new technological change, but may establish precisely what choices there are early in the process, and from this help to establish how new technology changes can best be handled.

Bibliography

Aglietta, M. (1979), *A Theory of Capitalist Regulation*, New Left Books, London.

Aitken, H.G.J. (1960), *Taylorism at Watertown Arsenal: Scientific Management in Action 1908-1915*, Harvard University Press, Massachusetts.

Armstrong, P. (1988), 'Labour and Monopoly Capital' in Hyman, R. and Streeck, W. (eds.), *New Technology and Industrial Relations*, Basil Blackwell, Oxford.

Arthurs, A. (1985), 'Egalitarianism in the Workplace.' in Hammond, V. (ed.), *Current Research in Management*, Francis Pinter, London.

Association of Professional, Executive, Clerical and Computer Staff (1985), *Job Design and New Technology*, APEX, London.

Association of Professional, Executive, Clerical and Computer Staff (Undated but issued 1987) *New Technology: Model Agreement, Notes for Guidance*, APEX.

Atkinson, J. (1985) - *Flexibility, Uncertainty and Manpower Management*, IMS Report no. 89, Institute of Manpower Studies, Brighton.

Atkinson, J. and Meager, N. (1986), 'Is Flexibility Just a Flash in the Pan', *Personnel Management*, September.

Babbage, C. (1835), 'On the Economy of Machinery and Manufacturers' in Davis, L.E. and Taylor, J.C. (eds.) (1979) *The Design of Jobs*, Goodyear, Santa Monica.

Baritz, L. (1960), *The Servants of Power*, Wesleyan University Press, Connecticut.

Batstone, E., Gourlay, S., Levie, H. and Moore, R. (1987), *New Technology and the Process of Labour Regulation,* Clarendon Press, Oxford.

Beechly, V. (1982), 'The Sexual Division of Labour and the Labour Process: A Critical Assessment of Braverman' in Woods, S. (ed.), *The Degradation of Work,* Hutchinson, London.

Bell, D. (1973), *The Coming of Post Industrial Society,* Basic Books, New York

Berg, I. (1981), 'Introduction' in Berg, I. (ed.), *Sociological Perspectives on Labor Markets,* Academic Press, New York.

Berg, I. (ed.) (1981), *Sociological Perspectives on Labour Markets,* Academic Press, New York.

Beynon, H. (1984), *Working for Ford,* Penguin, Harmondsworth.

Blackburn, P., Coombs, R. and Green, K. (1985), *Technology, Economic Growth and the Labour Process,* Macmillan Press, Basingstoke.

Blackburn, R.M. and Mann, M. (1979), *The Working Class in the Labour Market,* Macmillan, London.

Blauner, R. (1964), *Alienation and Freedom: The Factory Worker and His Industry,* University of Chicago Press, Chicago.

Bradley, K. and Hill, S. (1983), 'After Japan: The Quality Circle Transplant and Productive Efficiency', *Journal of Industrial Relations,* November, pp291-311.

Braverman, H. (1974), *Labor and Monopoly Capital: The Degradation of Work in the Twentieth Century,* Monthly Review Press, New York.

Briefs, U., Ciborra, C. and Schnieder, L. (eds.) (1983), *Systems design for, with and by the Users,* North Holland, Amsterdam.

Brown, W. (1983), 'Britains Unions: New Pressures and Shifting Loyalties', *Personnel Management,* October pp48-51.

Buchanan, D. A. (1986), 'Management Objectives in Technical Change' in Knights, D. and Wilmott, H., *Managing the Labour Process,* Gower, Aldershot.

Buchanan, D. and Boddy, D. (1986), *Managing New Technology,* Basil Blackwell, Oxford.

Burawoy, M (1979), *Manufacturing Consent,* University of Chicago Press, Chicago.

Burnes, B. (1988), 'New Technology and Job Design: the case of CNC', *New Technology Work and Employment,* Vol.3, No. 2, Autumn.

Cadbury, E. (1914), 'Some Principles of Industrial Organization', *Sociological Review,* 7, pp99-117.

Cavestro, W. (1989), 'Automation, New Technology and Work Content' in Wood, S. (ed.), *The Transformation of Work,* Unwin Hyman, London, pp219-234.

Chalmers, (1982), *What is This Thing Called Science?,* Open University, Milton Keynes.

Child, J. (1985) - 'Managerial Strategies New Technology and the Labour Process' in Knights, D., Willmot, H. and Collinson, D. (eds.), *Job Redesign: Critical Perspectives on the Labour Process*, Gower, Aldershot, pp.197-226.

Child, J. (1972), 'Organisational Structure, Environment and Performance: the Role of Strategic Choice' in Salaman, G. and Thompson, K. (eds.) (1973), *People and Organisations*, Longman, London, pp91-107.

Clark J. (1989), 'New Technology and Industrial Relations', *New Technology Work and Employment*, Vol.4, No.1, Spring, pp5-17.

Cockburn, C. (1983), *Brothers: Male Dominance and Technological Change*, Pluto Press, London.

Cockburn, C. (1984), 'New Technology in Print: Men's Work and Womens Chances' in Winch, G. (ed.) (1984), *Information Technology in Manufacturing Processes: Case Studies in Technological Change*, Rosendale, London, pp126-134.

Collard, R. (1981), 'The Quality Circle in Context', *Personnel Management*, September, pp26-30 and 51.

Collins, H.M. (1985), *Changing Order: Replication and Induction in Scientific Practice*, Sage Publications, London.

Cooley, M. (1981), *Architect or Bee?: The Human/Technology Relationship*, Langley Technical Services, Slough.

Coombes, R. (1978), 'Labour and Monopoly Capital', *New Left Review*, 107, January/February, pp79-96.

Coriat, B (1987), 'Information Technologies, Productivity and New Job Content: Skill as a competitive issue', *Brie Meeting on Comparative Productions*, 11-13 September, Berkeley.

Cressey, P., Di Martino, V., Bolle de Bal, M., Treu, T and Traynor, K. (1988), *Participation Review: A Review of Foundation Studies on Participation*, Office for Official Publications of the European Communities, Luxembourg.

Cressey, P (1990), 'Trends in Employee Participation and Industrial Democracy' in Russell, R. and Rus, V., *International Yearbook of Industrial Democracy*.

Davies, A. (1984), 'Management Union Participation During Microtechnological Change' in Warner, M. (ed.) (1984), *Microprocessors, Manpower and Society*, Gower, Aldershot, pp149-170.

Davis, L. E. (1955), 'Toward a Theory of Job Design', *Journal of Industrial Engineering,* vol. 1, No. 2, September/October.

Davis, L.E. (1971), 'Readying the Unready: Post Industrial Jobs', *California Management Review*, Vol. 14.

Davis, L.E. (1979), 'Future Directions' in Davis, L.E. and Taylor, J.C. (eds.), *The Design of Jobs*, Goodyear, Santa Monica.

Davis, L.E. (1979), 'Job Design: Historical Overview' in Davis, L.E. and Taylor, J.C. (eds.), *The Design of Jobs*, Goodyear, Santa Monica, pp.29-35.

Davis, L.E. and Taylor, J.C. (1976), 'Technology and Job Design', in Davis, L.E. and Taylor, J.C. (eds.) (1979), *The Design of Jobs*, Goodyear, Santa Monica.

Davis, L.E. and Taylor, J.C. (eds.) (1979), *The Design of Jobs*, Goodyear, Santa Monica.

Dickson, J.W. (1981), 'Participation as a Means of Organisational Control' *Journal of Management Science*, Vol. 18, No. 2, pp159-176.

Doeringer, P.B. and Piore, M.J. (1971), *Internal Labor Markets and Manpower Analysis,* Heath Lexington Books, Massachusetts.

Drucker, P. F. (1976), 'The Coming Rediscovery of Scientific Management', *Conference Board Record*, Vol. 13, June, pp23-27.

Edwards, R. (1979), *Contested Terrain: The Transformation of the Workplace in the Twentieth Century*, Heinemann, London.

Elger, T. (1984), 'Braverman, Capital Accumulation and Deskilling' in Wood, S. (ed.) (1984), *The Degradation of Work?* Hutchinson, London.

Emery, F.E. (ed.) (1969), *Systems Thinking*, Penguin, Harmonsworth.

Emery, F.E. and Trist, E.L. (1960), 'Sociotechnical Systems' in Emery, F.E. (ed.) (1969), *Systems Thinking*, Penguin, Harmondsworth, pp281-296.

Fox, A. (1974), *Beyond Contract: Work, Power and Trust Relations*, Faber and Faber, London.

Fox, A. (1985), *Man Mismanagement*, Hutchinson, London.

Friedman, A.L. (1977), *Industry and Labour: Class Struggle at Work and Monopoly Capitalism*, Macmillan, London.

Friedmann, G. (1961), *The Anatomy of Work*, Heinemann, London.

Gartman, D. (1978), Marx and the Labor Process: An Interpretation. in Essays on the Social Relations of Work and Labour, *The Insurgent Sociologist*, Vol. 8, No. 23, Fall, pp97-108.

Gennard, J. (1987), 'The NGA and the Impact of New Technology', *New Technology Work and Employment*, Vol. 1, pp126-141.

Gennard, J. and Dunn, S. (1983), 'The Impact of New Technology on the Structure and Organisation of Craft Unions in the Printing Industry', *British Journal of Industrial Relations*, Vol. 21, March, pp17-32.

Gill, C. (1985), *Work, Unemployment and the New Technology*, Polity Press, Cambridge.

Glenn, E. N. and Feldberg, R. L. (1979), 'Proletarianizing Clerical Work: Technology and Organizational Control in the Office' in Zimbalist, A. (ed.), *Case Studies on the Labor Process*, Monthly Review Press, New York.

Goldthorpe, J.H., Lockwood, D., Bechhofer, F. and Platt, J. (1968), *The Affluent Worker: Industrial Attitudes and Behaviour*, Cambridge University Press, Cambridge.

Gordon, G.M., Edwards, R. and Reich, M. (1982), *Segmented Work, Divided Workers: The historical transformation of labor in the United States*, Cambridge University Press, Cambridge.

Gorz, A. (ed.) (1976), *The Division of Labour: The Labour Process and Class Struggle in Modern Capitalism*, Harvester Press, Brighton.

Gospel, H. F. (1983), 'Managerial Structures and Strategies: An Introduction' in Gospel, H. F. and Littler, C. R. (eds.), *Managerial Strategies and Industrial Relations*, Heinemann, London.

Gospel, H. F. and Littler, C. R. (eds.), (1983), *Managerial Strategies and Industrial Relations*, Heinemann, London.

Goss, D. (1988), 'Diversity, Complexity and Technological Change: An Empirical Study of General Printing', *Sociology*, Vol. 22, No. 3, August.

Herding, R. (1972), *Job control and Union Structure: A Study on Plant Level Industrial Conflict in the United States with a Comparative Perspective on West Germany*, Rotterdam University Press, Rotterdam.

Hill, S. (1981), *Competition and Control at Work*, Heinemann, London.

Hirschhorn, L. (1984), *Beyond Mechanization: Work and Technology in a Post Industrial Age*, MIT, Massachusetts.

Hounshell, D.A. (1984), *From the American System to Mass Production 1800-1932: The Development of Manufacturing Technology in the United States*, The John Hopkins University Press, Baltimore.

Hyman, R. (1979), 'The Politics of Workplace Trade Unionism: Recent Tendencies and Some Problems for Theory', *Capital and Class*, Vol. 8, pp54-67.

Hyman, R. (1988), 'Flexible Specialization: Miracle or Myth' in Hyman, R. and Streeck, W. (eds.), *New Technology and Industrial Relations*, Basil Blackwell, Oxford.

Hyman, R. and Price, R. (eds.) (1983), *The New Working Class? White Collar Workers and their Organizations*, Macmillan, London.

Hyman, R. and Streeck, W. (eds.) (1988), *New Technology and Industrial Relations*, Basil Blackwell, Oxford.

Industrial Relations Review and Report (1985), 'Portsmouth and Sunderland Direct Input', *IRRR*, 352, 24 September.

Jones, B (1984), 'Destruction or Redistribution of Engineering Skills? The Case of Numerical Control' in Wood, S. (ed.) *The Degradation of Work?*, Hutchinson, London, pp179-199.

Jones, B. (1986), 'Cultures, Strategies and Technical Essentials: A Comparative View of Work and Flexible Production Technology' *International Workshop on New Technology and New Forms of Work Organisation*, Berlin (GDR), 9-12 November.

Jones, B. (1983), 'Technical Organisation and Political Constraints on System Redesign for Machinist Programming of NC Machine Tools' in Briefs, U., Ciborra, C. and Schnieder, L.

(eds.), *Systems Design for, with and by the User*. North Holland, Amsterdam.

Jones, B. (1985), 'Flexible Technologies and Inflexible Jobs: Impossible Dreams and Missed Opportunities', *World Congress on Human Factors in Automation*, Society of Manufacturing Engineers, Michigan.

Jones, B. (1988), 'Work and Flexible Automation in Britain: A Review of Developments and Possibilities', *Work, Employment and Society*, Vol. 2, No. 4, pp451-486.

Jones, B. (1990), 'New Production Technology and Work Roles: A Paradox of Flexibility versus Strategic Control', Loveridge, R. and Pitt, M (eds.), *The Strategic Management of Technological Innovation*, John Wiley and Sons, Chichester.

Jones, B. and Rose, M. (1985), 'The Trade Union Response to the Changing Job Structure of British Industry' in Spyropolous, G (ed.), *Trade Unions Today and Tomorrow Vol.2: Trade Unions in Changing Workplace*, Conference on 'Trade Unions in the Coming Decade' at the European Centre for Work and Society, Maastricht 20-22 November, Presses Interuniveritaires Europenes, Maastricht.

Jones, B. and Rose, M. (1986), 'Redividing Labour: Factory Politics and Work Reorganisation in the Current Industrial Transition' in Purcell, K., Wood, S., Waton, A. and Allen, S. (eds.), *The Changing Experience of Employment: Restructuring and Recession*, Macmillan, Basingstoke.

Jones, B. and Scott, P. (1987), 'Working the System: FMS in Britain and the USA', *New Technology, Work and Employment*, Vol. 2, No. 1, Spring.

Kamata, S. (1984), *Japan in the Passing Lane*, Unwin Paperbacks, London.

Kaplinsky, R. (1984), *Automation: the technology and society*, Longman, Harlow.

Kelley, M.R. (1989), 'Alternative forms of Work Under Programmable Automation' in Wood, S. (ed.), *The Transformation of Work*, Unwin Hyman, London.

Kelly, J. (1978), 'Understanding Taylorism: some comments', *British Journal of Sociology*, Vol. 29, No. 2, June.

Kelly, J.E. (1982), *Scientific Management, Job Redesign and Work Performance*, Academic Press, London.

Kelly, J.E. (1985), 'Managements Redesign of Work: Labour Process, Labour Markets and Product Markets' in Knights, D., Willmot, H. and Collinson, D. (eds.), *Job Redesign: Critical Perspectives on the Labour Process*, Gower: Aldershot, pp.30-51.

Knights, D. and Collinson, D. (1985), 'Redesigning Work on the Shopfloor: A Question of Control or Consent?' in Knights, D., Willmot, H. and Collinson, D. (eds.), *Job Redesign: Critical Perspectives on the Labour Process*, Gower: Aldershot, pp197-226.

Knights, D. and Wilmott, H. (eds.) (1986), *Managing the Labour Process*, Gower, Aldershot.

Knights, D., Willmot, H. and Collinson, D. (eds.) (1985), *Job Redesign: Critical Perspectives on the Labour Process*, Gower, Aldershot.

Kusterer, K.C. (1978), *Know How on the Job: The Important Working Knowledge of Unskilled Workers*, West View Press, Colorado.

Lawler, E.E. (1986), *High Involvement Management*, Jossey Bass, San Francisco.

Libetta, L. (1988), *Tacit Knowledge and the Computerisation of Skills*, Ph.D. Thesis, University of Bath.

Lintner, V. G., Pokorny, M. J., Woods, M. M., Blinkhorn, M. R. (1987), 'Trade Unions and Technological Change in the UK Mechanical Engineering Industry', *British Journal of Industrial Relations*, Vol. 25, No. 1, March, 1987.

Littler, C. R. (1978), 'Understanding Taylorism', *British Journal of Sociology*, Vol.29, No.2, June.

Littler, C. R. (1982), *The Development of the Labour Process in Capitalist Societies*, Heinemann, London.

Littler, C.R. (1983), 'A History of New Technology' in Winch, G. (ed.), *Information Technology in Manufacturing Processes: Case Studies in Technological Change*, Rosendale, London, pp135-144.

Littler, C.R. (1985), 'Taylorism, Fordism and Job Design' in Knights, D., Willmot, H. and Collinson, D. (eds.), *Job Redesign: Critical Perspectives on the Labour Process*, Gower, Aldershot, pp10-29.

Littler, C.R. and Salaman, G. (1984), *Class at Work: The Design, Allocation and Control of Jobs*, Batsford, London.

Littler, C.R. and Salaman, G. (1985), 'The Design of Jobs' in Littler, C.R. (ed.) (1985), *The Experience of Work* Gower, Aldershot, pp85-104.

Littler, C.R. (ed.) (1985), *The Experience of Work*, Gower, Aldershot.

Loveridge, R. and Pitt, M (eds.), (1990), *The Strategic Management of Technological Innovation*, John Wiley and Sons, Chichester.

Lupton, T. (ed.), (1984), *Proceedings of the 1st International Conference on Human Factors in Manufacturing*, 3-5 April, IFS Publications and North Holland, London.

Maier, C.S. (1970), 'Between Taylorism and Technocracy: European Ideologies and the Vision of Industrial Productivity in the 1920's', *Contemporary History*, Vol. 5, No. 2, pp27-62.

Mainwaring, T. (1984), 'The Extended Internal Labour Market', *Cambridge Journal of Economics*, Vol. 8, pp161-187.

Mainwaring, T. and Wood, S. (1985), 'The Ghost in the Labour Process' in Knights, D., Willmot, H. Collinson, D. (eds.) (1985), *Job Redesign: Critical Perspectives on the Labour Process*, Gower, Aldershot, pp171-196.

Martin, R. (1981), *New Technology and Industrial Relations in Fleet Street*. Oxford: Clarendon Press.

Martin, R. (1983), 'New Technology and Industrial Relations in Fleet Street: 'New technology will make it possible for managers to manage' in Warner, M. (ed.), *Microprocessors, Manpower and Society*, Gower, Aldershot, pp240-252.

McGregor, D. (1960), *The Human Side of Enterprise*, McGraw Hill, New York.

McLoughlin, I. and Clark, J. (1988), *Technological Change at Work*, Open University, Milton Keynes.

Mulkay, M. (1979), *Science and the Sociology of Knowledge*, George, Allen and Unwin, London.

Mumford, E. (1986), 'Helping Organisations Through Action Research: the Sociotechnical Approach', *Quality of Working Life*, Vol. 3, pp329-344,

Nadworny, M.J. (1955), *Scientific Management and the Unions 1900-32*, Harvard University Press, Massachusetts.

National Graphical Association (1984), 'The Way Forward: New Technology in the Provincial Newspaper Industry', (Council Policy Document), *NGA '82 Biennial Delegate Conference*, Blackpool, November.

National Graphical Association (Undated), *The Wolverhampton Express and Star: a briefing paper*, Published by the NGA.

National Union of Journalists/National Graphical Association, *NUJ/NGA Agreement: Direct Input in Provincial Newspapers*, (The Accord), Unpublished.

Nelson, D. F. (1980), *Frederick W. Taylor and the Rise of Scientific Management*, University of Wisconsin Press, Wisconsin.

Newspaper Society (1988), *Production Journal*, No.'s 130, January; 131, March; 132, May.

Newspaper Society (1987), *Newstime*, September and October editions.

Nichols, T. and Beynon, H. (1976), *Workers Divided*, Fontana, London.

Noble, D. F. (1979), 'Social Choice in Machine Design: The Case of Automatically Controlled Machine Tools' in Zimbalist, A (ed.), *Case Studies on the Labor Process*, Monthly Review Press, New York.

Palloix, C. (1976), 'The Labour Process: From Fordism to Neo Fordism', *CSE Pamphlet No.1 - The Labour Process and Class Strategies*, pp46-67.

Park, S. (1987), 'Smoothing the Long Road to Realism', *Newstime*, September, p24.

Park, S. (1987), 'The Direct Route to Success', *Newstime*, October, pp18-19.

Pearson Group (1986), *Annual Report*.

Phillips, A. (1983), Review of Cockburn, C. (1983), 'Brothers: Male Dominance and Technological Change', *Feminist Review*, Vol. 15, pp101-104.

Phillips, A. and Taylor, B. (1980), 'Sex and Skill: Notes towards a Feminist Economics', *Feminist Review* Vol. 6, pp79-88.

Pignon, D. and Querzola, J. (1976), 'Dictatorship and Democracy in Production' in Gorz, A. (ed.), *The Division of Labour: The Labour Process and Class Struggle in Modern Capitalism*, Harvester Press, Brighton.

Piore, M.J. (1968), 'The Impact of the Labor Market upon the Design and Selection of Production Techniques within the Manufacturing Plant' in *Quarterly Journal of Economics*, pp602-620.

Piore, M.J. (1980), *Birds of Passage: Migrant Labour and Industrial Societies*, Cambridge University Press, Cambridge.

Piore, M.J. (1986), 'Perspectives on Labour Market Flexibility' *Industrial Relations*, Vol. 25, No. 2, Spring, pp146-166.

Popper, K. (1966), *The Open Society and its Enemies*, Routledge and Kegan Paul, London.

Popper, K. (1972), *Conjectures and Refutations: the Growth of Scientific Knowledge*, Routledge and Kegan Paul, London.

Preece, D. A. (1986), 'Organisations, Flexibility and New Technology' in Voss, C. A. (ed.), *Managing Advanced Manufacturing Technology*, IFS Publications, Bedford.

Preece, D. A. (1987), 'New Technology and Job Design Lessons from the Print Industry?' *Employee Relations*, Vol. 9, No. 2.

Price, R. (1980), 'White Collar Unions: Growth, Character and Attitudes in the 1970's' in Hyman, R. and Price, R. (eds.), (1983), *The New Working Class? White Collar Workers and their Organizations*, Macmillan, London.

Price, R. (1988), 'Information Consultation and the Control of New Technology' in Hyman, R. and Streeck, W. (eds.), *New Technology and Industrial Relations*, Basil Blackwell, Oxford, pp249-262.

Purcell, K., Wood, S., Waton, A. and Allen, S. (eds.), (1986), *The Changing Experience of Employment: Restructuring and Recession*, Macmillan, Basingstoke.

Rainbird, H. (1988), 'New Technology, Training and Union Strategies' in Hyman, R. and Streeck, W. (eds.), *New Technology and Industrial Relations*, Basil Blackwell, Oxford, pp174-185.

Ramsay, H. (1985), 'What is Participation For? A Critical Evaluation of Labour Process Analyses of Job Reform' in Knights, D., Willmot, H. and Collinson, D. (eds.), *Job Redesign: Critical Perspectives on the Labour Process*, Gower, Aldershot, pp53-80.

Rose, M. (1978), *Industrial Behaviour: Theoretical Developments Since Taylor*, Penguin, Harmondsworth.

Rose, M. (1988), *Industrial Behaviour: Research and Control*, (Second edition), Penguin, Harmondsworth.

Rose, M. (1985), *Reworking the Work Ethic: Economic Values and Sociocultural Politics*, Batsford, London.

Rose, M. and Jones, B. (1985), 'Managerial Strategy and Trade Union Responses in Work Reorganisation Schemes', in Knights, D., Willmot, H. Collinson, D. (eds.), *Job Redesign: Critical Perspectives on the Labour Process*, Gower, Aldershot.

Rothwell, S.G. (1984), 'Company Employment Policies and the New Technology in Manufacturing and Service Sectors' in Warner, M. (ed.), *Microprocessors, Manpower and Society*, Gower, Aldershot.

Ryan, P. (1981), 'Segmentation, Duality and the Internal Labour Market' in Wilkinson, F., *The Dynamics of Labour Market Segmentation*, Academic Books, London, pp3-20.

Sabel, C.F. (1982), *Work and Politics*, Cambridge University Press, Cambridge.

Sabel, C.F. and Zietlin, T. (1985), 'Historical Alternatives to Mass Production: Politics, Markets and Technology in Nineteenth Century Industrialisation', *Past and Present*, 108, August, pp133-76.

Salaman, G. (1986), *Working*, Tavistock Publications, London.

Salaman, G. and Thompson, K. (eds.) (1973), *People and Organisations*, Longman, London.

Sell, R. (1986), 'The Politics of Workplace Participation', *Personnel Management*, June, pp34-37.

Shaiken, H. (1985), *Work Transformed: Automation and Labor in the Computer Age*, Holt, Rinehart and Winston, New York.

Smith, A. (1776), *The Wealth of Nations*, republished 1974, Penguin, Harmondsworth.

Smith, A. (1986), 'Flexibility at the FT' *Industrial Society*, Vol. 68, December, pp22-24.

Smith, P. (1988), 'The Impact of Trade Unionism and the Market in a Regional Newspaper' *Industrial Relations Journal*, Vol. 19, No. 3, Autumn, pp214-221.

Smith, P. and Morton, G. (1990), 'A Change of Heart: Union Exclusion in the Provincial Newspaper Sector', *Work Employment and Society*, Vol. 4, No. 1, March, pp105-124.

Smith, R. and Quinlan, T (1982), 'Croydon Advertiser Group', *Work Research Unit*.

Sorensen, A.B. and Kalleberg, A.L. (1981), 'An Outline of a Theory of the Matching of Persons to Jobs' in Berg, I. (ed.), *Sociological Perspectives on Labour Markets*, Academic Press, New York, pp49-74.

Sorge, A. and Streeck, W. (1988), 'Industrial Relations and Technical Change: The case for an Extended Perspective' in Hyman, R. and Streeck, W. (eds.), *New Technology and Industrial Relations*, Basil Blackwell, Oxford.

Sorge, A., Hartman, G., Warner, M. and Nicholas, I. (1983), *Microelectronics and Manpower in Manufacturing: Applications of Numerical Control in Great Britain and West Germany*, Gower, Aldershot.

Spyropolous, G. (ed.) (1985), *Trade Unions Today and Tomorrow Vol.2: Trade Unions in Changing Workplace*, from the conference on 'Trade Unions in the Coming Decade' at the European Centre for Work and Society, Maastricht, 20-22 November. Presses Interuniveritaires Europenes: Maastricht.

Storey, J, (1987), 'The Management of New Office Technology: Choice, Control and Social Structure in the Insurance Industry', *Journal of Management Studies*, Vol 24, No. 1, January.

Storey, J. (1985), 'The Means of Management Control', *Sociology*, Vol. 19, No. 2, May, pp193-211.

TASS (1978), *New Technology: A guide for negotiators*, AUEW/TASS, Richmond.

Taylor, F.W. (1908), 'Shop Management' in Taylor, F.W. (1947), *Scientific Management*, Harper and Brothers, New York.

Taylor, F.W. (1911), 'The Principles of Scientific Management' in Taylor, F.W. (1947), *Scientific Management*, Harper and Brothers, New York.

Taylor, F.W. (1912), 'Taylor's Testimony Before the Special House Committee' in Taylor, F.W. (1947), *Scientific Management*, Harper and Brothers, New York.

Taylor, F.W. (1947), *Scientific Management*, Harper and Brothers, New York.

Tanner, F. (1981), 'Stamp Out the Parasites', *Print* (NGA Journal), February.

Thompson, L. (1985), *New Office Technology: People, Work structure and the Process of Change*, Work Research Unit Occasional Paper No. 34., April.

Thompson, C.B. (1914), 'Some Principles of Industrial Organization' *Sociological Review*, Vol. 7, pp315-327.

Trades Union Congress (1979), *Employment and Technology*, TUC, London.

Trist, E.L. and Bamforth, K.W. (1951), 'Some Psychological and Social Consequences of the Longwall Method of Coal Getting', *Human Relations*, Vol. 4.

Thyer, G. (1988), *Computer Numerical Control of Machine Tools*, Heinemann, Oxford.

Voss, C. A. (1986), *Managing Advanced Manufacturing Technology*, IFS Publications, Bedford.

Wainwright, J. and Francis, A. (1986), *Office Automation, Organisation and the Nature of Work*, Gower, Aldershot.

Walby, S. (1989), 'Flexibility and the Changing Sexual Division of Labour' in Wood, S. (ed.) (1989), *The Transformation of Work*, Unwin Hyman, London, pp127-140.

Wall, T. and Kemp, N. (1987), 'The Nature and Implications of Advanced Manufacturing Technology: Introduction' in Wall, T.D., Clegg, C.W. and Kemp, N.J. (eds.), *The Human Side of Advanced Manufacturing Technology*, John Wiley and Sons, Chichester, pp1-14.

Wall, T.D., Clegg, C.W. and Kemp, N.J. (eds.) (1987), *The Human Side of Advanced Manufacturing Technology*, John Wiley and Sons, Chichester.

Wallace, T.A.H. and Whitehall, F.B. (1984), 'Some Industrial Relations Aspects of New Technology in the Machine Shop Environment' in Lupton, T. (ed.), *Proceedings of the 1st International Conference on Human Factors in Manufacturing*, 3-5 April, IFS Publications and North Holland, London, pp209-226

Warner, M. (ed.) (1984), *Microprocessors, Manpower and Society*, Gower, Aldershot.

Warr, P. and Wall, J. (1975), 'History of Work Concepts' in Davis, L.E. and Taylor, J.C. (eds.) (1979), *The Design of Jobs*, Goodyear, Santa Monica, pp29-35.

Whitaker, R. (1979), 'Scientific Management Theory as Political Ideology', *Studies in Political Economy*, 2, pp75-108.

White, G.C. (1984), *Employee Involvement in Work Redesign*, WRU Occasional Paper No. 29, February.

White, G.C. (1983), *Redesign of Work Organisations - its impact on Supervisors*, WRU Occasional Paper No. 26, August.

White, G.C. (1977), *Job Design and Individual Differences*, WRU Occasional Paper No. 9, September.

White, G.C. (1982), *Technological Changes and Employment: A review of some implications and union policies*, WRU Occasional Paper No. 22, July.

Wilkins, P. M. (1990), *New Technology and the Job Definition Dilemma*, PhD Thesis, University of Bath, Bath.

Wilkinson, B. (1983), *The Shopfloor Politics of New Technology*, Heinemann, London.

Wilkinson, F. (1981), *The Dynamics of Labour Market Segmentation*, Academic Books, London.

Williams, R. and Steward, F. (1985), 'Technology Agreements in Great Britain: A Survey 1977-83' *Industrial Relations Journal*, Vol. 16, No. 3, Autumn, pp58-73.

Wilson, F.M. and Buchanan, D.A. (1988), 'New Technology in the Engineering Industry: Cases of Control and Constraint', *Work, Employment and Society*, Vol. 2, No. 3, September, pp366-380.

Winch, G. (ed.) (1983), *Information Technology in Manufacturing Processes: Case Studies in Technological Change*, Rosendale, London, pp135-144.

Wood, S. (ed.) (1989), *The Transformation of Work*, Unwin Hyman: London.

Wood, S. and Kelly, J. (1982), 'Taylorism, Responsible Autonomy and Management Strategy' in Wood, S., *The Degradation of Work?* Heinemann, London, pp74-89.

Wood, S. ed. (1984), *The Degradation of Work?*, Hutchinson, London.

Woodward, J. (1966), *Industrial Organization: Theory and Practice*, Oxford University Press, Oxford.

Work Research Unit (1982), *Meeting the Challenge of Change: Case Studies*, WRU, London.
Zimbalist, A. (1975), 'The Limits of Work Humanization', *Review of Radical Political Economics*, Vol. 7, no. 2, pp50-59.
Zimbalist, A. (1979a), 'Introduction' in Zimbalist, A. (ed.), *Case Studies on the Labor Process*, Monthly Review Press, New York, ppxi-xxiv
Zimbalist, A. (1979b), 'Technology and the Labor Process in the Printing Industry' in Zimbalist, A. (ed.), *Case Studies on the Labor Process*, Monthly Review Press, New York.
Zimbalist, A. (ed.), (1979), *Case Studies on the Labor Process*, Monthly Review Press, New York.